LEAN SIX SIGMA for ENGINEERS and MANAGERS

With Applied Case Studies

LEAN SIX SIGMA for ENGINEERS and MANAGERS

With Applied Case Studies

Matthew John Franchetti

CRC Press
Taylor & Francis Group
Boca Raton London New York

CRC Press is an imprint of the
Taylor & Francis Group, an **informa** business

CRC Press
Taylor & Francis Group
6000 Broken Sound Parkway NW, Suite 300
Boca Raton, FL 33487-2742

Printed on acid-free paper
Version Date: 20150306

International Standard Book Number-13: 978-1-4822-4352-9 (Hardback)

Library of Congress Cataloging-in-Publication Data

Franchetti, Matthew J.
Lean Six Sigma for engineers and managers : with applied case studies / author, Matthew John Franchetti.
pages cm
Includes bibliographical references and index.
ISBN 978-1-4822-4352-9 (hardback)
1. Six sigma (Quality control standard) I. Title.

TS156.17.S59F73 2015
658.4'01--dc23 2015006884

Visit the Taylor & Francis Web site at
http://www.taylorandfrancis.com

and the CRC Press Web site at
http://www.crcpress.com

Contents

Preface

Lean Six Sigma (LSS) is an emerging process and product management strategy that combines the quality and customer satisfaction benefits of Six Sigma with the cost reduction benefits of Lean manufacturing. The need for LSS has been brought on by increased global competition, sustained financial crises, and increased consumer expectations for higher quality and lower costs. Technologically complex products and processes combined with global supply chains have intensified the need for LSS. The benefits of focusing on LSS by individual organizations can lead to enhanced economic performance, strong levels of customer satisfaction, and higher market shares.

Lean Six Sigma for Engineers and Managers: With Applied Case Studies provides an up-to-date source of technical information relating to current and potential LSS practice and implementation. This book provides a detailed framework and reference material for LSS and is presented in three sections:

Section I: Provides the definition, benefits, and background information regarding LSS

Section II: Discusses LSS strategies and provides the general framework to implement LSS

Section III: Provides several case studies in various service and manufacturing organizations

This book serves as both a practical and technical guide to understand and implement LSS fully for any organization, from manufacturing to service facilities. It is based on concepts related to total quality management, data analysis, and statistical process control. The LSS process provided in this book has been applied and refined over the past 10 years on more than 20 LSS projects around the globe. This book incorporates several case studies to aid in the implementation of such programs. I am confident that this book will provide a contribution to the Six Sigma and Lean manufacturing field and emphasize the need for continued research as we strive for increased competiveness in the rapidly evolving global market.

About the Author

Matthew John Franchetti, PhD, is an assistant professor of mechanical, industrial, and manufacturing engineering and the director of undergraduate studies of the Mechanical and Industrial Engineering Programs at The University of Toledo. Dr. Franchetti earned his PhD in 2003 and MBA in 2000 from The University of Toledo. He has worked as an industrial engineer and technical manager for the US Postal Service and has extensive consulting experience in the automotive industry. Dr. Franchetti is a certified Six Sigma black belt from the American Society for Quality (ASQ) and has consulting and research experience with more than 25 companies across the country. He has published more than 60 books, articles, and conference proceedings in the field of industrial engineering.

Section I

Why Lean Six Sigma?

1

Definition and Motivation for Lean Six Sigma

1.1 Introduction

It is not necessary to change. Survival is not mandatory.

W. Edwards Deming, 2000 [1]

The ability for an organization to adapt quickly to changing market conditions while simultaneously delivering high levels of customer service at superior quality and low cost are no longer differentiators in the competitive marketplace: they are minimum requirements to survive in a global business environment. Thomas L. Friedman, author of *The World Is Flat*, pioneered discussions regarding the challenges and opportunities related to the merging of world economies and the need for organizations to adapt; as a result of advances in information technology and logistics, the playing field is leveling [2]. China, Europe, India, and the United States are now competing for the same customers and operating in a global economy; all are also influenced by conditions in the integrated economies. Friedman points out that this was painfully evident in the 2008 economic crisis where the entire global economy at one time acted totally in synch as a result of the mortgage crisis in the United States [3].

In this evolving, fast-paced, and highly competitive worldwide marketplace, agility and the ability to deliver high quality at a low cost is crucial to an organization's success and survival. Lean Six Sigma is a process management system that allows organizations to accomplish this through the combination of Lean, which focuses on cost reductions with Six Sigma, which focuses on quality improvements. The Association of Manufacturing Excellence (AME) defines Lean manufacturing as a customer-based production practice that focuses on the creation of value for the end customer and considers the expenditure of resources for anything other than customer value to be wasteful, and thus a target for elimination [4]. The American Society for Quality (ASQ) defines Six Sigma as a fact-based, data-driven philosophy of quality improvement that values defect prevention over defect detection; it drives customer satisfaction and bottom-line results by reducing variation and waste, thereby promoting a competitive advantage [5]. Six Sigma is a combination of a management philosophy, a set of improvement tools, and a methodology rolled into one system [5]. Lean Six Sigma combines Lean and Six Sigma methodologies into one integrated process and management system that simultaneously works to reduce operating costs and improve quality for the end customer.

General Electric (GE) attributes Six Sigma implementation to driving more than $10 billion of benefits at GE and driving up operating margins from 14.8% to 18.9% in four years [6]. Following are two specific examples of success at GE [6]:

- At GE Capital, over 24% of incoming calls in the mortgage division were going to voicemail. A Six Sigma team found that one branch had a near perfect percentage of answered calls. After the team analyzed its system, process flows, equipment, physical layout, and staffing, this branch model was duplicated everywhere and the company met customer calls with a live GE person on the first try 99.9% of the time.
- In GE Plastics, they were denied bidding on contracts to provide Sony CD-ROMs due to quality issues. GE implemented a Six Sigma improvement plan and reduced defect rates from 1.2% to 0.00001% and subsequently won the Sony contract.

Many analytical models have been developed to help companies achieve the goals of cost reductions and quality improvements. One such method is presented in this book. This method has been applied to over 50 companies and has achieved demonstrated quality and bottom-line economic results. The model hinges on the business maxim, "If you can't measure it, you can't manage it." The ultimate goals of the method presented in this book allow organizations to:

- Increase customer satisfaction through improved quality and consistent delivery.
- Increase business competitiveness through reduced operating costs.
- Accelerate the ability of an organization to adapt and thrive in a changing business environment.
- Improve corporate images as the organizations excel beyond competitors.

The overarching goals of this book are to empower companies and organizations to outshine competitors in terms of cost and quality and to adapt to change. Many organizations are able to increase profits for segments of their operations by reducing costs, raw material purchases, and other operating costs. However, many companies do not have the capability to perform comprehensive Lean Six Sigma initiatives due to time constraints and lack of knowledge in the field. This book provides a detailed framework and reference material for implementing Lean Six Sigma and it is broken into the three sections:

- Section I provides definitions, benefits, and background information regarding Lean Six Sigma.
- Section II discusses Lean Six Sigma strategies and provides the general framework for implementation and sustained success.
- Section III features 11 case studies in various industries.

1.2 Definitions

In order to better understand Lean Six Sigma, it is first necessary to discuss the terms and definitions that relate to the process. In the process management field, people often have differing expectations upon hearing many of the common terms. To compound the problem, finding universal definitions for these terms can be challenging as many companies and government agencies create their own designations, often using combinations of technical and operational components. This section discusses these key terms and definitions as they relate to the topics covered in this book.

Six Sigma is a comprehensive and flexible system for achieving, sustaining, and maximizing business success [7]. It is uniquely driven by a close understanding of customer needs; disciplined use of facts, data, and statistical analysis; and diligent attention to managing, improving, and reinventing business processes [6]. Six Sigma strives to improve quality, productivity, and bottom-line financial performance. The Six Sigma approach provides a methodology to achieve these successes and provides a system on which to base any improvement initiative, including energy and greenhouse gas minimization programs. Six Sigma relies heavily on data, facts, and the use of statistical tools to study whether an improvement has been made. These are powerful statistical tools for conducting experiments, comparing data, and providing important information about a process to find the causes of problems and draw conclusions. Six Sigma methodologies are broken into six fundamentals:

- Define products or services.
- Know the stakeholders and customers and their critical needs.
- Identify processes, methods, and systems to meet stakeholders' critical needs.

- Establish a process of doing work consistently.
- Error-proof process and eliminate waste.
- Measure and analyze performance.

To achieve these fundamental goals, Six Sigma uses the five-step DMAIC methodology described below:

DEFINE

Define and clarify the project goal and timeline with the focus placed on the end customers.

Establish the team, including senior leadership.

MEASURE

Define the current state and current processes, including the development of a process flow map to baseline the system and to identify any bottlenecks and establish current financial and quality performance.

Collect and display data including task time, resources required, and process statistics.

ANALYZE

Determine process capability and speed utilizing statistical tools and charts.

Determine sources of variation and subsequent time bottlenecks.

Identify and quantify value-added and non-value-added activities.

IMPROVE

Generate ideas, including:

Methods to improve quality and reduce costs.

Process changes.

Conduct experiments and validate improved processes.

Develop action plans and standard operating procedures.

CONTROL

Develop control plan.

Monitor performance.

Mistake-proof processes.

Six Sigma has three overall targeted solutions: process improvement, process design/redesign, and process management. Process improvement refers to a strategy of correcting an issue while leaving the core structure of the process intact. Process redesign/re-engineering is a strategy to overhaul the process completely, including the core structure. Process management refers to integrating Six Sigma methods into everyday business, including:

- Processes are documented and managed "end-to-end," and responsibility has been assigned in such a way as to ensure cross-functional management of critical processes.
- Customer requirements are clearly defined and regularly updated.
- Measures of outputs, process activities, and inputs are thorough and meaningful.
- Managers and associates use the measures and process knowledge to assess performance in real time and take action to address problems and opportunities.
- Process improvement and process design/redesign are used to raise the company's levels of performance, competitiveness, and profitability constantly.

The Lean Six Sigma implementation procedure is segmented into four phases as displayed in Figure 1.1 which provides an overview of the procedure.

1.2.1 Acronyms and Symbolism

Below is a list of commonly used acronyms in the Lean Six Sigma field:

ANOVA—Analysis of Variance
BRM—Business Risk Management
CI—Continuous Improvement
COPQ—Cost of Poor Quality
CTC—Critical to Customer
CTQ—Critical to Quality
DFMEA—Design Failure Mode and Effect Analysis
DFSS—Design for Six Sigma
DMAIC—Define, Measure, Analyze, Improve, and Control
DOE—Design of Experiment
DPMO—Defects Per Million Opportunity
ECO—Engineering Change Order

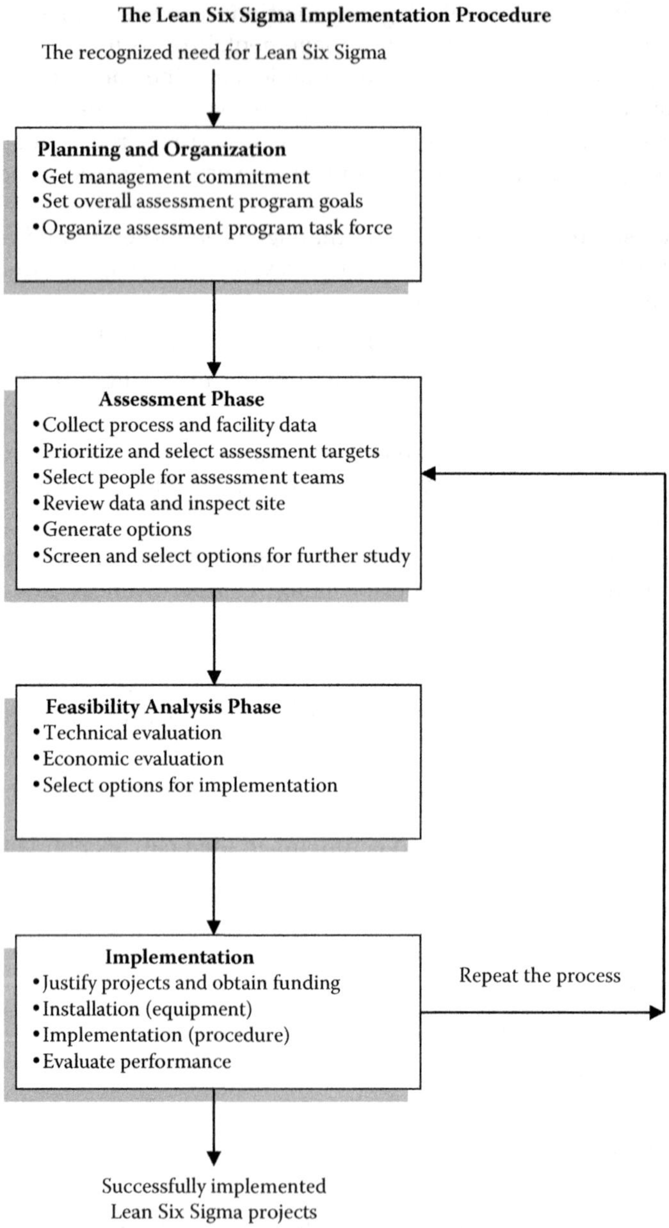

FIGURE 1.1
The Lean Six Sigma implementation process.

FAST—Function Analysis System Technique
FIFO—First In First Out
GQTS—Global Quality Tracking System
GRPI—Goals, Roles, Processes, and Interpersonal Relationships
JIT—Just In Time
KBI—Key Business Issue
LIFO—Last In First Out
LLB—Lean Level of Buffering
OEE—Overall Equipment Effectiveness
OEM—Original Equipment Manufacturer
PCE—Process Cycle Efficiency
PPM—Parts Per Million
QFD—Quality Function Deployment
ROI—Return on Investment
RTY—Rolled Throughput Yield
SIPOC—Suppliers, Inputs, Process, Output, and Customers
SMART—Specific, Measurable, Achievable, Relevant, and Time-bound
SOP—Standard Operating Procedures
SPC—Statistical Process Control
TOC—Theory of Constraints
TPS—Toyota Production System
TQM—Total Quality Management
VOB—Voice of Business
VOC—Voice of Customer
VOE—Voice of the Employee
VSM—Value Stream Mapping
WIP—Work in Process

1.2.2 Units

In the field of Lean Six Sigma, measurements are typically taken in parts per million opportunity (PPMO) for a given product, activity, or process and return on investment. Table 1.1 displays common metric system units for other measurements typically taken during the Lean Six Sigma process.

TABLE 1.1

Commonly Used Metric System Units and Symbols in Lean Six Sigma

Quantity Measured	Unit	Symbol
Mass	metric ton	t
Temperature	degree Celsius	°C
Volume	cubic meter	m^3
Power	kilowatt	kW
Energy	kilowatt-hour	kW·h
Time	second	s
Length	kilometer	km

1.3 System Approach to Lean Six Sigma Implementation

The Lean Six Sigma process presented in this book is based on the systems approach. The systems approach is a problem-solving philosophy that focuses on a holistic view of an organization by analyzing the linkages and interactions between the elements that comprise the entire system. A system can be defined as group of interacting, interrelated, or interdependent elements forming a complex whole coordinated to achieve a stated purpose or goal. The systems approach emphasizes that the best method to understand a problem is to understand the individual parts in relation to the whole. From a macro view, a system is comprised of inputs, processes, and outputs all revolving around accomplishing a given goal or goals. The definition and clear understanding of this goal is critical to defining the system in terms of its processes, required inputs, and desired outputs. For example, there will be very different systems for an organization that produces solar cells versus an organization that provides food services.

The key benefit of the systems approach to Lean Six Sigma is that it addresses the problem from a business standpoint, consistently focusing on the organization's goals, and confronting the problem at every stage of the supply chain. Traditional manufacturing approaches only tend to address quality at the end of the process, in a retroactive manner. Many organizations also manage cost and quality as compartmentalized "problems" that are managed separately from its core processes. The central issue with this traditional approach is that by focusing on these individual outcomes, overall system optimization cannot be achieved. The systems approach addresses issues at all phases of the supply chain from raw material procurement to the design of quality-driven processes. Figure 1.2 provides an overview of the system as it relates to business processes and Lean Six Sigma implementation.

Defining the terms of a system will provide additional insights into the interactions and relationships as they process improvement. The starting

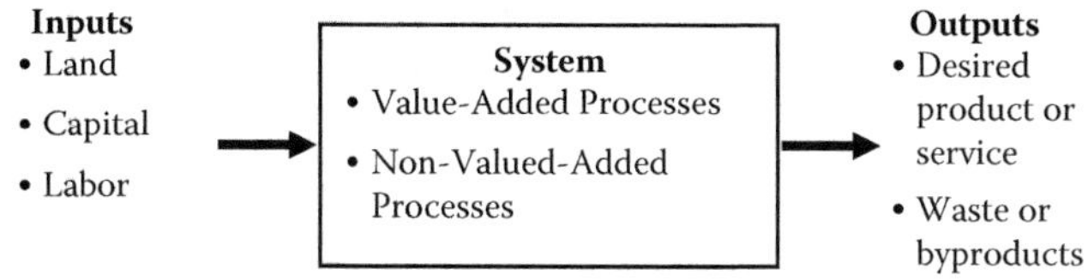

FIGURE 1.2
The systems approach overview.

point of any system is to define its goal or purpose clearly. Once the goal is defined, the required inputs, outputs, and process designs can be determined concurrently, focusing on the system goals. For example, suppose an organization sets a goal to manufacture 30,000 television sets per year at a rate of return of 14%. Now the system can be defined in terms of processes, inputs, and outputs. The inputs, also known as the factors of production, are land, capital, and labor:

- *Land or natural resources:* Naturally occurring goods such as water, air, soil, or minerals used in the creation of products.
- *Labor:* Human effort used in production which also includes technical and marketing expertise.
- *Capital:* Human-made goods (or means of production) that are used in the production of other goods. These include machinery, tools, and buildings.

From the television manufacturing example, the inputs would include the raw materials, such as the electronics, plastics, and metals. Considering these raw materials, the processes would be designed concurrently. Attention would be given to value-added and non-value-added activities. The process design will answer the following questions:

- What is to be produced?
- How are the products to be produced?
- When are the products to be produced?
- How much of each product will be produced?
- For how long will the products be produced?
- Where are the products to be produced?

The answers to these questions are obtained from product, process, and schedule design and are not independently and sequentially determined, but developed concurrently. A clear vision is critical and the success of a firm is dependent on having an efficient production system. The product design phase defines which products will be produced and provides detailed drawings of the part. A quality deployment function (QDF) is generally applied to accomplish this. A QDF is an organized approach to identify customer needs and translate the needs to product characteristics, process design, and the tolerances required. The process design phase examines these issues:

- How is the product to be produced?
- Make or buy decision
- Process selection
- Equipment selection
- Process time determination

The desired outputs of the design phase are:

- Processes
- Equipment
- Raw materials required
- Formal documents:
  - Route sheet
  - Assembly chart
  - Operation process chart
  - Precedence diagram

The schedule design phase examines the following issues:

- How much to produce? (lot sizing)
- When to produce? (production scheduling)
- How long to produce? (market forecasts)

The determination of the outputs of a system seems very straightforward at first examination, as it should be defined in the goal. The systems approach is an integral concept to the successful implementation of Lean Six Sigma and serves as the basis for the solution methodologies. This holistic approach considers the entire supply chain and addresses cost and quality use issues at the earliest phases possible. When promoting Lean Six Sigma to decision makers, it is critical to relate the benefits to the financial performance of the organization. Creating this win–win situation in a manager's eyes will significantly enhance the likelihood that the Lean Six Sigma recommendations will be implemented. The remaining chapters of this book provide a detailed road map for Lean Six Sigma implementation and case studies.

References

1. Deming, E.W. (2000). *Out of the Crisis*. Cambridge, MA: MIT Press.
2. Friedman, T.L. (2007). *The World Is Flat,* 3rd edition, New York: Picador.
3. Friedman T.L. and Mandelbaum, M. (2012). *That Used to Be Us: How America Fell Behind in the World It Invented and How We Can Come Back.* New York: Picador.
4. Womack, J.P. and Jones, D.T. (1996). *Lean Thinking: Banish Waste and Create Wealth in Your Corporation.* New York: Simon & Schuster.
5. Benbow, D.W. and Kubiak, T.M. (2005). *The Certified Six Sigma Black Belt Handbook.* Milwaukee: ASQ Quality Press.
6. Pande, P., Neuman, R., and Cavanagh, R. (2000). *The Six Sigma Way: How GE, Motorola, and Other Top Companies Are Honing Their Performance.* New York: McGraw Hill.
7. Everett Decision Systems. (2011). *Lean Six Sigma.* Retrieved from https://www.everettdecisions.com/Lean_Six_Sigma.html

2

History and Fundamentals

2.1 Introduction

Lean and Six Sigma are an evolution of traditional industrial engineering practices that date back to the late 1890s. When Fredrick Winslow Taylor, the father of scientific management, first used a stopwatch in 1891 to measure, improve, and standardize the paper mill processes in the plant that he managed, Lean and Six Sigma were begun [1]. Taylor's mantra for the scientific management system that he created was, "It is only through enforced standardization of methods, enforced adoption of the best implements and working conditions, and enforced cooperation that this faster work can be assured. And the duty of enforcing the adoption of standards and enforcing this cooperation rests with management alone." This could still be applied today as organizations explore methods to improve quality and lower costs [2]. The remainder of this chapter provides a brief history of Lean Six Sigma to provide the necessary context related to the improvement methodology.

2.2 Brief History of Lean Six Sigma

The following is a brief history of Lean Six Sigma; Lean Six Sigma should be considered an evolution of modern process management practices that date back to the mid-1760s when the Industrial Revolution led to fundamental changes as the economy shifted from an agricultural base to an industrial base in England [3]. The Industrial Revolution later followed into the United States in the mid-1820s [3] and led to several evolutionary milestones powered by advances in information technology that culminated in the Lean Six Sigma movement of the current day.

2.2.1 Industrial Revolution (1760s)

In the mid-1700s, the world economies underwent a major shift from an agricultural base to an industrial base, as societies became larger and more organized. The Industrial Revolution was a period in the late eighteenth and early nineteenth centuries when major changes in agriculture, manufacturing, and transportation had a profound effect on socioeconomic and cultural conditions around the world [3]. It started with the mechanization of the textile industries, the development of iron-making techniques, and the increased use of refined coal. Trade expansion was enabled by the introduction of canals, improved roads, and railways. The introduction of steam power (fueled primarily by coal) and powered machinery (mainly in textile manufacturing) underpinned the dramatic increases in production capacity. The development of all-metal machine tools in the first two decades of the nineteenth century facilitated the manufacture of more production machines for manufacturing in other industries. The effects spread throughout Western Europe and North America during the nineteenth century, eventually affecting most of the world [4].

2.2.2 Scientific Management (1890–1940)

Many consider Fredrick Winslow Taylor the father of Lean manufacturing, and he planted the seeds for what is now Six Sigma by emphasizing data collection and analysis to make management decisions in the early 1890s. Scientific management is defined as the management of a business, industry, or economy according to principles of efficiency derived from experiments in methods of work and production, primarily with the use of time-and-motion studies [1]. Taylor first introduced scientific management in the paper mill and still industries and paved the way for modern-day mass production [2]. Scientific management enforces the rigorous collection of data, continual improvement to find the "one best way" to conduct every manufacturing operation, and the need for management to strictly monitor and enforce standard operating practices. The scientific management methodology is the forefather of modern-day Lean manufacturing.

2.2.3 Toyota Production System/Lean Manufacturing (1945–1950)

The Toyota Production System (TPS) is an integrated sociotechnical system developed by the Toyota Motor Corporation shortly after the end of World War II to allow the company to become more competitive and cost effective [5]. The main objectives of TPS are to design out overburdening (muri) and inconsistency (mura), and to eliminate the seven wastes (muda) [5]. The TPS focuses on the end customer to create processes that drive end value and consistent delivery for the customer. The TPS is the precursor to just-in-time and Lean manufacturing and initially focused on the automobile industry, but may be applied to other industries as well [5].

2.2.4 Information Revolution (1950–1980)

Upon the end of World War II, the world economies began to shift back toward normal production; in addition, the "game-changing" electronic age began with the invention of the computer [6]. The information revolution served as a catalyst to allow organizations to collect, process, and disseminate data quickly within and between businesses. Advanced statistical analyses were now possible in short periods of time as managers and engineers no longer relied on hand calculations for analysis and improvement. As the world shifted to "knowledge economies" and the usage of computers became more widespread, the door opened to modern quality control processes [6].

2.2.5 Total Quality Management (1950–1985)

Total quality management (TQM) is a management approach to long-term success through customer satisfaction in which all members of an organization participate in improving processes, products, services, and the culture in which they work [7]. TQM emphasizes statistical analysis and statistical process control to monitor and improve the quality of business processes and products. The value of TQM concepts first gained reknown in Japan after World War II, when W. Edwards Deming and Joseph Duran, two American statisticians, traveled there to implement statistical quality control concepts during the war rebuilding effort. TQM emphasizes customer focus, total employee involvement, process-centered approaches, integrated systems, strategic and systematic approaches, continual improvement, fact-based decision making, and strong communication [7]. Six Sigma eventually evolved from the TQM methodology.

2.2.6 ISO 9000 (1985 and Beyond)

ISO (International Organization for Standardization) is the world's largest developer of voluntary international standards [8]. In 1987, the ISO 9000 Quality Standard was published; the ISO 9000 family of standards is related to quality management systems designed to help organizations ensure that they meet the needs of customers and other stakeholders while meeting statutory and regulatory requirements related to a product [9]. In addition to developing the standard, the ISO provides third-party certification to designate that an organization has an acceptable quality management system in place that is effectively utilized. The third-party ISO 9000 certification is meaningful in that it creates new and sustained business opportunities for organizations that have achieved the certification as many view that it designates a seal of approval that an organization has implemented an effective quality management system.

2.2.7 Six Sigma (1990 and Beyond)

Six Sigma is a quality improvement methodology with the ideal outcome of reaching zero defects. It can be considered the next evolution of total quality management. In the late 1980s Motorola began implementing the Six Sigma methodology under CEO Bob Galvin, and the company became a quality leader in industry [10]. General Electric followed shortly after and achieved similar successes; GE successes with Six Sigma led to a rapid proliferation of the Six Sigma methodology in industry.

Six Sigma primarily functions in three different ways: as a metric, management system, and methodology. As a metric, Six Sigma is a scale of a company's quality level; for an organization to have a "Six Sigma level" of quality, the company must only have 3.4 defects per one million opportunities (DPMO) or less. As a management system, Six Sigma pulls the concepts of Six Sigma into a corporate business strategy. As a methodology, Six Sigma is represented by the DMAIC model [10]. DMAIC is the acronym for define, measure, analyze, improve, and control. The major advances of the Six Sigma methodology over total quality management relate to:

1. The establishment of stretch goals to achieve zero defects
2. The creation of the DMAIC process improvement cycle
3. The heavy use of statistics and data to make management decisions and reduce variation in processes

In addition to these advances, the Six Sigma field has also designated various levels of personal expertise in applying Six Sigma: green belt (novice to Six Sigma with some experience in Six Sigma projects), black belt (expert in applying and leading Six Sigma improvement projects), and master black belts (organization leaders who oversee all Six Sigma efforts and execution plans within an organization). These personal designations are typically granted by a third party, such as a professional society or educational institute or may be granted via a certification process within the organization applying Six Sigma. More details regarding Six Sigma are provided in Section II of this book.

2.2.8 Lean Six Sigma (2005 and Beyond)

Lean Six Sigma is a managerial concept combining Lean and Six Sigma and focuses on simultaneous quality improvements and cost reductions for an organization, product, or process. The Lean Six Sigma concept was first introduced in 2002 by Michael George in the book titled *Lean Six Sigma: Combining Six Sigma with Lean Production Speed* [11]. The combination of Lean and Six Sigma laid the foundation to combat the largest criticisms of Lean and Six Sigma as stand-alone methodologies, specifically for Lean that it was not concerned with the quality of the product and for Six Sigma that it was not concerned with cost reductions. Before the introduction of Lean Six Sigma, many managers viewed quality improvements and cost reduction initiatives as mutually exclusive ventures; the Lean Six Sigma concept opened the door to the fact that both can and should be implemented simultaneously. The remainder of this book provides a detailed road map and several case studies to implement Lean Six Sigma to assist organizations in achieving dual quality and cost improvements.

2.3 Quality and Lean Leaders

Throughout modern history, several individuals have pioneered quality and Lean efforts and have made a significant impact on the body of knowledge in their respective fields. These individuals were ahead of their time and provided paradigm shifts in the field. Like many on the cutting edge who push the boundaries of a given field, their ideas initially were not fully accepted by their peers, reinforcing the fact that change can be difficult for organizations. Through demonstrated results, their ideas gained widespread acceptance and are today considered standards in the quality and Lean fields. By understanding their backgrounds and efforts, insight can be gained into the next big paradigm shifts in the Lean and Six Sigma fields. The following details these individuals and their contributions.

2.3.1 Fredrick Winslow Taylor (1856–1915)

Fredrick Winslow Taylor was an American mechanical engineer and is considered the father of scientific management [1]. In 1878, he was a machine shop laborer at Midvale Steel Works who was promoted into management and held the belief that the shop floor workers were not highly efficient and not working hard enough; as a result, he believed that the company was losing a significant amount of money. These pressures were the catalyst for his scientific management concept and his goal to create standard procedures based on time studies for every operation under his control to ensure efficiency and cost control. His efforts marked the origin of "Lean manufacturing" and the concepts related to value streams, worker/machine utilization, productivity analysis, and what would later become mass production. In 1898, he accepted a position at Bethlehem Steel to improve capacity and lower costs [1]. Many at the company felt his management concepts were too aggressive and he was forced to leave in 1901, underscoring the resistance to change for many organizations very early in American history.

2.3.2 Henry Ford (1863–1947)

Henry Ford was the founder of Ford Motor Company in 1903 and is credited with further developing the assembly line and ushering in mass production as a widely accepted manufacturing system [12]. Many of his management concepts fit closely with Taylor's scientific management philosophy; Ford modified these concepts for the automobile industry on a very large scale. In the early 1900s, he had a very innovative management philosophy for his time: to pay high wages to workers to ensure high productivity and quality while systematically lowering production costs [12]. In fact, he was paying double the labor rates to his production workers versus the industry average at the time to bolster employee morale and productivity. His management philosophy took Taylor's scientific management style of "Do it this way in this amount of time—or you're fired," combined with an incentive program of high wages. Ford's higher wages also had the benefit of creating a workforce that could afford his product, the Model-T, increasing his sales further [12].

2.3.3 W. Edwards Deming (1900–1993)

W. Edwards Deming was an American statistician, professor, and consultant who led quality efforts in postwar Japan in the 1950s [13]. His major contribution to the field related to the use of statistics and quality control processes to minimize and monitor defects. He pioneered quality efforts in Japan and is regarded as a national hero for his contributions that transformed Japan into a producer of top-quality goods. He faced a great deal of adversity early in his career and resistance to change for his philosophies; initially in the 1940s, his quality control ideas were not accepted by US manufacturers. When he attempted to promote these concepts he met with much resistance and limited buy-in to his concepts. In the 1950s he was requested to travel to Japan to assist with the rebuilding efforts and his quality control concepts gained widespread acceptance [13].

2.3.4 Joseph M. Juran (1904–2008)

Joseph Juran was an electrical engineer by training, and along with Deming he pioneered quality improvement efforts in Japan after World War II. His passion for quality control and improvement was fostered by an early position; he joined Western Electric's Hawthorne Works to troubleshoot customer issues in the complaint department [14]. Most of his contributions to the field relate to the use of statistics for process improvement. He is credited with creating the Pareto principle, acceptance sampling, and statistical process control [14].

2.3.5 Kiichiro Toyoda (1894–1952)

Kiichiro Toyoda was a Japanese entrepreneur who brought Toyoda Loom Works into automobile manufacturing and created what would eventually become Toyota Motor Corporation [15]. His decision to transform Toyoda into an automobile company laid the foundation for what is now the

Toyota Production System (TPS) and created the world's largest automobile manufacturer. TPS is considered the benchmark for Lean manufacturing and is a large reason Toyota has been so dominant in the automobile industry over the past four decades [15].

2.3.6 Philip B. Crosby (1926–2004)

Philip Crosby was a quality leader who promoted the concept of zero defects. He stressed the cost benefits of zero defects, and in his most popular book, *Quality Is Free*, he demonstrated that the money a company invests to improve quality and achieve zero defects will result in positive cash flows related to reduced rework, reduced internal/external failures, and reduced appraisals/prevention costs [16]. He also emphasized that one cannot "inspect in" quality retroactively, that it must be built into the process proactively to eliminate defects at all stages of the process and eliminate the need for inspection [16]. His work laid the foundation for what is now Six Sigma by focusing a stretch goal of zero defects and integrating quality measures throughout the entire process or system.

2.3.7 Armand V. Feigenbaum (1922–2014)

Armand Feigenbaum is credited with developing the total quality management (TQM) philosophy while working for General Electric as an operations manager [17]. His concepts transformed the statistical process control techniques developed by Deming, Juran, and Crosby into a company-wide movement that engaged all employees in quality improvement. TQM represented the first company-wide quality improvement strategy and in many ways, was the predecessor of Six Sigma.

2.3.8 Kaoru Ishikawa (1915–1989)

Kaoru Ishikawa was an engineering professor at the University of Tokyo and he developed the cause and effect diagrams (also known as fishbone diagrams) that are used to identify root causes for process improvement [18]. A fishbone diagram is displayed in Figure 2.1. He promoted and enhanced the statistical process control concepts developed by Deming and Juran in postwar Japan, and enhanced these efforts by developing quality circles to incorporate all levels of the organization in the problem-solving process.

2.3.9 Genichi Taguchi (1924–2012)

Genichi Taguchi was an engineer in postwar Japan for the Electrical Communications Laboratory of the Nippon Telegraph and Telephone Corporation and is credited with developing new statistical methods to measure the cost of quality [19]. He created what is now known as the Taguchi

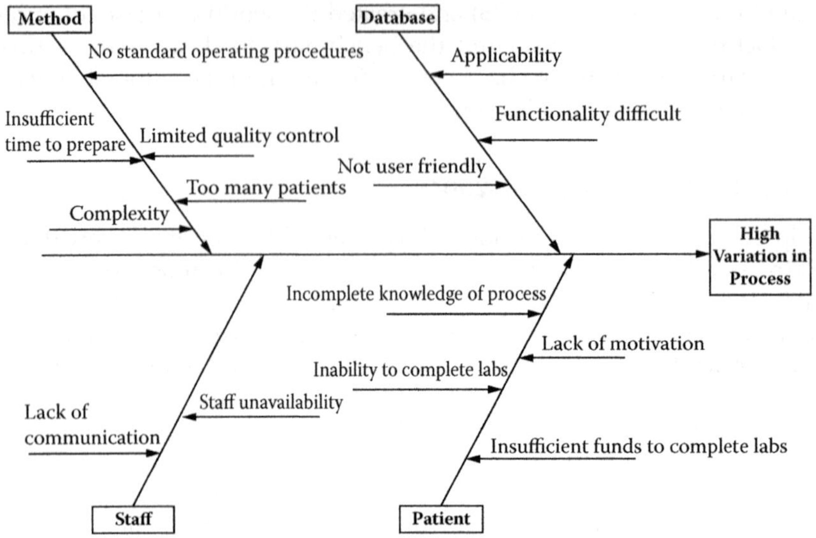

FIGURE 2.1
Fishbone diagram example.

loss function, which is a statistical method to measure the financial loss to society from poor quality. His methods emphasized the use of design of experiments (DOE) to better understand the process and material parameters that maximized quality of the end product and robustness in design.

2.3.10 Mikel Harry (1955–)

Mikel Harry has been widely recognized and cited in many publications as the principal architect of Six Sigma and a leading authority within the field [20]. He led efforts at Motorola to improve quality in the late 1980s by creating and implementing what is now known as the Six Sigma philosophy. His efforts were recognized by competing companies and in the mid-1990s General Electric (GE), under the leadership of CEO Jack Welch, implemented Six Sigma with tremendous success [21]. Motorola and GE's quality and financial successes during the 1990s led to the proliferation of the Six Sigma philosophy to numerous organizations in the United States and the world.

2.4 Quality Standards and Systems

2.4.1 ISO 9001

The ISO 9000 family addresses various aspects of quality management; the standards provide guidance and tools for companies and organizations that want to ensure that their products and services consistently meet customer's requirements, and that quality is consistently improved [22]. The standards were first published by the nonprofit International Standard Organization in 1987 and represented the first external quality certification for businesses.

2.4.2 QS 9000

QS 9000 is a quality standard developed jointly by General Motors, Chrysler, and Ford for the automotive industry and suppliers in 1994 [23]. It is similar to ISO 9000, but specifically focuses on the automotive industry. Many automotive manufacturers require that their suppliers are QS 9000 certified before any contracts are finalized. QS9000 ensures that the entire automotive supply chain for a specific vehicle (including all components and processes) have a documented process to monitor, track, and correct quality issues that may arise. This certification allows for the rapid tracking and resolution of quality issues that may arise at any point in the supply chain from any supplier.

2.4.3 ISO 14000

The ISO 14000 family addresses various aspects of environmental management and provides practical tools for companies and organizations looking to identify and control their environmental impact and constantly improve their environmental performance [24]. ISO 14000 was introduced in 1994 and focuses on environmental quality management such as lifecycle analysis, carbon footprints, and waste management. The ISO 14000 standard and certification ensure that an organization is properly tracking and handling waste, monitoring emissions, and reducing its environmental impacts.

2.5 Relevant Organizations and Individual Certifications

2.5.1 American Society for Quality (ASQ)

ASQ, formed in the 1960s, is a global community of people dedicated to quality who share the ideas and tools that make our world work better. ASQ provides the global quality community with training, professional certifications, and knowledge to a network of members [25]. This organization is closely tied in with Six Sigma philosophies and offers several well-recognized quality-related certifications for individuals including [26]:

- *Certified Quality Inspector (CQI):* The certified quality inspector is an inspector who, in support of and under the direction of quality engineers, supervisors, or technicians, can use the proven techniques included in the body of knowledge. Under professional direction, the quality inspector evaluates hardware documentation, performs laboratory procedures, inspects products, measures process performance, records data, and prepares formal reports.

- *Certified Quality Technician (CQT):* The certified quality technician is a paraprofessional who, in support of and under the direction of quality engineers or supervisors, analyzes and solves quality problems, prepares inspection plans and instructions, selects sampling plan applications, prepares procedures, trains inspectors, performs audits, analyzes quality costs and other quality data, and applies fundamental statistical methods for process control.
- *Certified Reliability Engineer (CRE):* The certified reliability engineer is a professional who understands the principles of performance evaluation and prediction to improve product/systems safety, reliability, and maintainability. This body of knowledge (BOK) and applied technologies include, but are not limited to, design review and control; prediction, estimation, and apportionment methodology; failure mode effects and analysis; the planning, operation, and analysis of reliability testing and field failures, including mathematical modeling; understanding human factors in reliability; and the ability to develop and administer reliability information systems for failure analysis, design and performance improvement, and reliability program management over the entire product life cycle.
- *Certified Manager of Quality/Organizational Excellence (CMQ/OE):* The certified manager of quality/organizational excellence is a professional who leads and champions process-improvement initiatives—everywhere from small businesses to multinational corporations—that can have regional or global focus in a variety of service and industrial settings. A certified manager of quality/organizational excellence facilitates and leads team efforts to establish and monitor customer/supplier relations, support strategic planning and deployment initiatives, and help develop measurement systems to determine organizational improvement. The certified manager of quality/organizational excellence should be able to motivate and evaluate staff, manage projects and human resources, analyze financial situations, determine and evaluate risk, and employ knowledge management tools and techniques in resolving organizational challenges.
- *Certified HACCP Auditor (CHA):* The certified hazard analysis and critical control points (HACCP) auditor is a professional who understands the standards and principles of auditing a HACCP-based (or process-safety) system. A HACCP auditor uses various tools and techniques to examine, question, evaluate, and report on that system's adequacy and deficiencies. The HACCP auditor analyzes all

elements of the system and reports on how well it adheres to the criteria for management and control of process safety.

- *Certified Biomedical Auditor (CBA):* The certified biomedical auditor is a professional who understands the principles of standards, regulations, directives, and guidance for auditing a biomedical system while using various tools and techniques to examine, question, evaluate, and report on that system's adequacy and deficiencies. A biomedical auditor analyzes all elements of the system and reports on how well it adheres to the criteria for management and control of process safety.
- *Certified Six Sigma Black Belt (CSSBB):* The certified Six Sigma black belt is a professional who can explain Six Sigma philosophies and principles, including supporting systems and tools. A black belt should demonstrate team leadership, understand team dynamics, and assign team member roles and responsibilities. Black belts have a thorough understanding of all aspects of the DMAIC model in accordance with Six Sigma principles. They have basic knowledge of Lean enterprise concepts, are able to identify non-value-added elements and activities, and are able to use specific tools.
- *Certified Pharmaceutical GMP Professional (CPGP):* The certified Pharmaceutical Goods and Manufacturing Processes (GMP) professional understands the GMP principles as regulated and guided by national and international agencies for the pharmaceutical industry. This covers finished human and veterinary drugs and biologics, ectoparasitacides, and dietary supplements (alternatively called nutraceuticals) where regulated as drug products, as well as their component raw materials (includes active pharmaceutical ingredients [APIs] and excipients), and packaging and labeling operations.
- *Certified Software Quality Engineer (CSQE):* The certified software quality engineer understands software quality development and implementation; software inspection, testing, verification, and validation; and implements software development and maintenance processes and methods.
- *Certified Quality Auditor (CQA):* The certified quality auditor is a professional who understands the standards and principles of auditing and the auditing techniques of examining, questioning, evaluating, and reporting to determine a quality system's adequacy and deficiencies. The certified quality auditor analyzes all elements of a quality system and judges its degree of adherence to the criteria of industrial management and quality evaluation and control systems.
- *Certified Quality Engineer (CQE):* The certified quality engineer is a professional who understands the principles of product and service quality evaluation and control. This body of knowledge and applied technologies include, but are not limited to, development

and operation of quality control systems, application and analysis of testing and inspection procedures, the ability to use metrology and statistical methods to diagnose and correct improper quality control practices, an understanding of human factors and motivation, facility with quality cost concepts and techniques, and the knowledge and ability to develop and administer management information systems and to audit quality systems for deficiency identification and correction.

- *Certified Quality Improvement Associate (CQIA):* The certified quality improvement associate has a basic knowledge of quality tools and their uses and is involved in quality improvement projects, but doesn't necessarily come from a traditional quality area.
- *Certified Calibration Technician (CCT):* The certified calibration technician tests, calibrates, maintains and repairs electrical, mechanical, electromechanical, analytical, and electronic measuring, recording, and indicating instruments and equipment for conformance to established standards.
- *Certified Quality Process Analyst (CQPA):* The certified quality process analyst is a paraprofessional who, in support of and under the direction of quality engineers or supervisors, analyzes and solves quality problems and is involved in quality improvement projects. A certified quality process analyst may be a recent graduate or someone with work experience who wants to demonstrate his or her knowledge of quality tools and processes.
- *Certified Six Sigma Green Belt (CSSGB):* The Six Sigma green belt operates in support of or under the supervision of a Six Sigma black belt, analyzes and solves quality problems, and is involved in quality improvement projects. A green belt is someone with at least three years of work experience who wants to demonstrate his or her knowledge of Six Sigma tools and processes.
- *Certified Master Black Belt (MBB):* The ASQ master black belt (MBB) certification is a mark of career excellence and aimed at individuals who possess exceptional expertise and knowledge of current industry practice. Master black belts have outstanding leadership ability, are innovative, and demonstrate a strong commitment to the practice and advancement of quality and improvement. Obtaining an ASQ MBB is acceptance and recognition from your peers.

2.5.2 Society of Manufacturing Engineers (SME)

Founded in 1932 as a nonprofit, the aim of SME is to connect all those who are passionate about making things that improve the world; the vision of SME is to enhance progress, prosperity, and strong communities through manufacturing [27]. SME is closely aligned with Lean manufacturing concepts and offers one Lean manufacturing certification [28]:

- *Lean Certification:* The Lean certification program comprises three levels—Bronze, Silver, and Gold—to represent the growing achievement by Lean practitioners. Individuals begin their journey by earning the Lean Bronze Certification (LBC), focusing on the tactical level. As their careers progress, candidates build upon their professional knowledge and skill set while advancing to the Silver (integrative) and Gold (strategic) certification levels.

2.5.3 Institute of Industrial Engineers (IIE)

The IIE was founded in 1948 and is an international nonprofit association that provides leadership for the application, education, training, research, and development of industrial engineering; including Lean and Six Sigma concepts [29]. The IIE offers several Lean and Six Sigma certificates for individuals [30]:

- *Engineering Management Certificate Program:* This program provides an individual with the basic management skills necessary for leading teams, departments, and organizations.
- *Healthcare Management Engineering Certificate Program:* This series of seminars provides an individual with the basics of the industrial engineering toolbox as applied to healthcare and provides an introduction to the fundamental management skills required in this industry.
- *Industrial Engineering Professional Skills Certificate Program:* This series of seminars presents the basics of the industrial engineering toolbox and introduces an individual to the fundamental management skills needed to take on demanding tasks in the industry.
- *Lean Enterprise Certificate Program:* This series of seminars is for individuals new to Lean who need to understand and apply the basic tools for Lean transformation. Courses will assist individuals in all types of businesses, including service, manufacturing, healthcare, government, finance, and education.
- *Six Sigma Certificate Program:* The IIE Six Sigma certificate program requires the completion of a total of 19 days of training delivered in three seminars. Completion of the certificate program will earn an individual both green belt and black belt certificates as well as the Six Sigma certificate.
- *Lean Six Sigma Certificate Program:* The IIE Lean Six Sigma certificate program requires the completion of a total of 25 days of training

delivered in two seminars. Completion of the certificate program will earn an individual both Lean Six Sigma green belt and Lean Six Sigma black belt certificates as well as the Lean Six Sigma certificate. This program is offered for both general business and healthcare applications.

- *Enterprise Risk Manager Certificate:* This certificate program is based on and addresses the new families of ISO, ANSI, and NIST standards dealing with risk management and responds to new federal regulatory requirements dealing with public safety.

2.6 Quality and Lean Awards

Over the past 60 years, several quality and Lean awards have been established for individuals or organizations that made key contributions to the Six Sigma or Lean field. Following is a brief list of several of the major awards given annually.

2.6.1 Malcolm Baldrige National Quality Award

The Baldrige Award was established by the US National Institute of Standards and Technology (NIST) in 1987 and recognizes US companies to (1) identify and recognize role-model businesses, (2) establish criteria for evaluating improvement efforts, and (3) disseminate and share best practices [31]. The Malcolm Baldrige National Quality Award is the highest level of national recognition for performance excellence that a US organization can receive. The US Congress originally authorized the Baldrige Award to include manufacturing, service, and small business organizations; Congress expanded eligibility to education and health care organizations in 1998 [32]. To receive the Baldrige Award, an organization must have a role-model organizational management system that ensures continuous improvement in the delivery of products or services, demonstrates efficient and effective operations, and provides a way of engaging and responding to customers and other stakeholders [32]. The criteria are organized into seven categories: leadership; strategic planning; customer focus; measurement, analysis, and knowledge management; workforce focus; operations focus; and results [33]. Previous recipients include Honeywell, Purina Pet Foods, Boeing, Caterpillar, and Motorola.

2.6.2 Eli Whitney Productivity Award

The Eli Whitney Productivity Award is presented annually by the Society of Manufacturing Engineers (SME) for distinguished accomplishments in improving capability within the broad concept of orderly production [34]. This award started in 1957 and recognizes advances and contributions related to Lean manufacturing and productivity.

2.7 Relevant Laws and Regulations

In terms of Lean manufacturing, productivity improvements, and quality initiatives several regulations govern the field. The Fair Labor Standards Act (FLSA) in the United States establishes a minimum wage, overtime pay, recordkeeping, and youth employment standards affecting employees in the private sector and in federal, state, and local governments; overtime pay at a rate not less than one and one-half times the regular rate of pay is required after 40 hours of work in a workweek. In terms of minimizing labor costs, the FLSA plays a key role in balancing fairness to employees and corporate profitability. Liability laws and litigation in the United States govern physical and property damages related to product failure or misuse. Six Sigma efforts can dramatically improve product quality and reduce liability issues.

References

1. Kanigel, R. (2005). *The One Best Way: Frederick Winslow Taylor and the Enigma of Efficiency*. Cambridge, MA: MIT Press.
2. Taylor, F.W. (1911). *The Principles of Scientific Management*. New York: Harper & Brothers,
3. Ashton, T.S. (1998). *The Industrial Revolution*. Oxford University Press: New York.
4. Deane, P.M. (1980). *The First Industrial Revolution*. London. Cambridge University Press.
5. Ohno, T. (1988). *Toyota Production System: Beyond Large-Scale Production*. New York: Productivity Press.
6. Davis, D., Miller, G.J., and Russell, A. (2005). *Information Revolution: Using the Information Evolution Model to Grow Your Business*. New York: Wiley.
7. Goetsch, D. and Davis. S. (2012). *Quality Management for Organizational Excellence: Introduction to Total Quality*. New York. Prentice Hall.
8. ISO. (2014). *The Story of ISO*. Geneva, Switzerland: http://www.iso.org/iso/home/about/the_iso_story.htm
9. Poksinska, B., Dahlgaard, J.J., and Antoni, M. (2002). The state of ISO 9000 certification: A study of Swedish organizations. *TQM Magazine*, 14(5): 297.
10. Pyzdek, T. (2009).*The Six Sigma Handbook*, Third Edition. New York: McGraw-Hill Professional.

11. Geroge, Michael L. (2002). *Lean Six Sigma: Combining Six Sigma Quality with Lean Production Speed.* New York: McGraw Hill.
12. Ford, H.,. Bakken, J.K., Bodek, N., and Crowther, S. (1988).*Today and Tomorrow—Special Edition of Ford's 1926 Classic.* New York. Productivity Press.
13. Deming, W.E. (2000). *The New Economics for Industry, Government, Education.* Cambridge, MA: MIT Press.
14. Juran, J.M. (2004). *Architect of Quality: The Autobiography of Dr. Joseph M. Juran.* New York: McGraw-Hill.
15. Kazuo, W. and Yui, T. (2002). *Courage and Change the Life of Kiichiro Toyoda.* Tokyo: Toyota Motor Corporation.
16. Crosby, P.B. (1979). *Quality Is Free: The Art of Making Quality Certain: How to Manage Quality—So That It Becomes A Source of Profit for Your Business.* New York: McGraw-Hill.
17. Feigenbaum, A.V. (1961). *Total Quality Control.* New York: McGraw-Hill.
18. Ishikawa, K. (1988). *What is Total Quality Control? The Japanese Way.* New York: Prentice Hall.
19. Taguchi, G., Subir Chowdhury, S., and Wu, Y. (1966). *Taguchi's Quality Engineering Handbook.* New York: Prentice Hall.
20. Harry, M.J. (1991). *The Vision of Six Sigma: Mathematical Constructs Related to Process Centering.* Illinois: Motorola University Press.
21. Slater, R. (1998). *Jack Welch & the G.E. Way: Management Insights and Leadership Secrets of the Legendary CEO: Management Insights and Leadership Secrets of the Legendary CEO.* New York: McGraw Hill.
22. ISO 9000. (2014). *Management System Standards.* http://www.iso.org/iso/iso_9000
23. Smith, R.M. (1998).*The QS-9000 Answer Book.* Chico, CA: Patton Press.
24. ISO 14000. (2014.) *Environmental Management.* http://www.iso.org/iso/home/standards/management-standards/iso14000.htm
25. The American Society for Quality. (2014). *Who We Are.* http://asq.org/about-asq/who-we-are/index.html
26. The American Society for Quality. (2014). *ASQ Quality Certifications.* http://cert.asq.org/certification/control/certifications
27. The Society of Manufacturing Engineers. (2014). SME History. https://www.sme.org/sme-history/
28. The Society of Manufacturing Engineers. (2014). *Lean Certification.* http://cert.asq.org/certification/control/certifications
29. The Institute of Industrial Engineers. (2014). *About IIE.* http://www.iienet2.org/
30. The Institute of Industrial Engineers. (2014). *Certificate Programs.* http://www.iienet2.org/IIETrainingCenter/Details.asp
31. The National Institute for Standards and Technology. (2014). http://www.nist.gov/
32. The National Institute for Standards and Technology. (2014). *About the Baldrige Award.* http://www.nist.gov/baldrige/about/baldrige_faqs.cfm
33. The National Institute for Standards and Technology. (2014). *The Baldrige Criteria for Performance Excellence.* http://www.nist.gov/baldrige/publications/criteria.cfm
34. The Society of Manufacturing Engineers. (2014). *International Honors and Awards.* https://www.sme.org/honor-awards/

3

Benefits of Lean Six Sigma

3.1 Introduction

The purpose of a project, plan, or initiative (Lean Six Sigma or otherwise) is to achieve measurable results that can be tied into a predetermined goal or future state. These results or benefits are critical in determining the feasibility and acceptance of a project or initiative such as Lean Six Sigma for an organization. These benefits are also the key selling points used when promoting Lean Six Sigma to stakeholders and decision makers. The benefits of Lean Six Sigma can be separated into five areas:

- Increased profitability and cost reduction
- Improved quality and customer satisfaction
- Enhanced management strategy, flexibility, and agility
- Heightened environmental protection
- Improved employee skill development, morale, and job satisfaction

Ideally, an organization would like to create a situation where multiple benefits can be realized from a single project or initiative. This synergistic approach allows for the creation of win–win situations when applied appropriately using the system approach discussed in this book. Specifically, the company will realize cost benefits and improved quality, the environment will be better protected, and the stakeholders of the organization (including employees) often gain a sense of well-being and satisfaction with the organization. This chapter discusses in greater detail these benefits and includes examples that may be used to promote Lean Six Sigma to decision makers.

3.2 Profitability and Cost Reduction

One of the primary benefits of Lean Six Sigma is associated with improved corporate profitability and operational cost reductions. The Lean and Six

Sigma aspects of the overall methodology address process improvement from different angles: Lean focuses on cost reductions (lower operational costs by reducing waste) and Six Sigma focuses on improved profitability (higher quality and customer satisfaction that leads to increased sales). Following is a list of cost benefits possible from the proper implementation of Lean Six Sigma:

- Bottom-line cost savings
- Reduction in process cycle times
- High external quality ratings
- Increased customer satisfaction
- Less space requirements
- Improved profitability
- Improved customer loyalty
- Fewer defects and scrap, nonconformance (reduced cost of internal and external quality failures)
- Safer work environment and less clutter

In 2001, Ford Motor Company reported saving $300 million ($52 million to the bottom line) through the implementation of Lean Six Sigma [1]. Ford implemented over 1,000 Six Sigma projects to achieve the cost savings and reported a two-point increase in customer satisfaction in 2001 [1]. Other companies have also achieved significant savings from the implementation of Six Sigma and Lean Six Sigma [2]:

- "GE saved $12 billion over five years and added $1 to its earnings per share. Honeywell (AlliedSignal) recorded more than $800 million in savings; additionally, GE produces annual benefits of over $2.5 billion across the organization from Six Sigma."
- "Motorola reduced manufacturing costs by $1.4 billion from 1987–1994."
- "Six Sigma reportedly saved Motorola $15 billion over the last 11 years."

Nonprofit and military operations have also achieved cost benefits from the implementation of Lean Six Sigma. In 2007 the US Army announced savings of nearly $2 billion with the implementation of Six Sigma principles across units. Successful initiatives optimized meal scheduling in dining areas, communication through various levels of service, task management, and fuel recycling [3].

3.3 Quality and Customer Satisfaction

Linked closely with increased profitability and operational cost reductions, improved quality, and increased customer satisfaction are other key benefits of Lean Six Sigma implementation. These benefits work together and feed one another; for example, when customer satisfaction improves, sales increase, and when quality improves, costs are reduced and profitability increases. Following is a brief list of quality and customer satisfaction benefits derived from the implementation of Lean Six Sigma:

- Improved quality of the product or service as perceived by the customer (for both internal and external customers)
- Higher quality as rated by external reviewers
- Fewer defects, scrap rates, and customer returns

In terms of quality improvements, an electronics manufacturer reported quality improvements from a 0.8% defect rate for a high-volume circuitboard to a defect rate of 0.004% through the implementation of Lean Six Sigma [4].

3.4 Management Strategy, Flexibility, and Agility

The tangible and readily measureable benefits of Lean Six Sigma such as improved profitability and customer satisfaction receive much media attention, but very strong, long-term, and deeper organizational benefits can be realized related to strategy enhancements. The Lean Six Sigma philosophy provides a common data-driven problem-solving approach that permeates all levels of the organization. Specifically, Lean Six Sigma can:

- Provide a common language throughout the organization.
- Enhance corporate and process problem solving.
- Improve and streamline supply chain management.
- Create effective partnerships and communication channels.
- Enhance agility to recognize and change rapidly to market conditions.
- Enhance flexibility.
- Improve employee motivation.
- Enhance time management.
- Reduce inventory levels.

3.5 Environmental Protection

Lean manufacturing has a strong link to green engineering. The focus of Lean is to eliminate all forms of waste in any system. The key element of Lean manufacturing that serves as a stepping stone to green engineering is the detailed process analysis involved that centers on waste identification and reduction. Once the wastes, including sources and amounts, are identified and the process is fully understood, steps can be taken to improve the overall process. By keeping environmental impact in the forefront during the Lean process analysis and design, initiatives can be developed to make a company more "green." Such focus can lead to reduced waste levels, smaller facility footprints, and reduced scrap rates.

3.6 Employee Skill Development and Job Satisfaction

Lean Six Sigma also develops the analytical skill sets and soft skills of employees. Six Sigma is based on the application of statistics to identify and eliminate defects and quality issues; teaching these skill sets to employees enhances their decision making and makes them better employees. Specifically, Six Sigma can aid in:

- Development of staff skills
- Improved knowledge and skills
- Improved employee morale
- Ability to use a wide range of tools and techniques

References

1. Six Sigma Success Stories. (2014). *Go Lean Six Sigma.* http://www.goleansixsigma.com/lean-six-sigma-success-stories-in-automotive-and-heavy-equipment-industries
2. iSix Sigma. (2014.) *Six Sigma Costs and Savings.* http://www.isixsigma.com/implementation/financial-analysis/six-sigma-costs-and-savings/
3. The Small Business Chronicle. (2014). *Advantages of Six Sigma.* http://smallbusiness.chron.com/advantages-six-sigma-43184.html
4. Rantamäki, J., Tiainen, E.L., and Kässi, T. (2013). A case of implementing SPC in a pulp mill. *International Journal of Lean Six Sigma*, 4: 321–337.

Section II

Lean Six Sigma Methodologies and Strategies

4

The Goal: Improving Quality and Reducing Costs

4.1 Introduction

The overarching goal of Lean Six Sigma is to simultaneously improve quality and reduce costs for an organization. It provides a common philosophy and process management approach for organizations to improve their performance. Oftentimes, quality improvements and cost reductions are viewed as contradictory goals. For example, many managers believe that if an organization places effort and resources toward quality improvement, operational costs must rise in terms of additional labor, tools, facility space, and software purchases to monitor and improve quality. On the other side of the coin, managers also believe that if an organization reduces costs, fewer resources will be available to monitor, improve, and control quality. As Philip Crosby pointed out in the late 1970s with his book, *Quality Is Free*, these are major misconceptions [1]. The misconceptions are still prevalent today in many service, manufacturing, and nonprofit organizations. This chapter discusses how quality truly is "free" (and in many cases can reduce costs and improve profitability) and how cost reduction and quality improvement goals work in harmony.

4.2 Cost of Poor Quality

The costs of poor quality are costs that occur when a product or service does not meet customer standards; in other words, if a company did not produce any defective products (100% perfect quality) these costs would not exist. An example of this is a fast-food drive-through that gives a customer the wrong order. If a customer ordered a double cheeseburger, small fries, and diet soda and they are given a chicken sandwich, no fries, and a root beer, what costs (or potential costs) does the fast-food restaurant incur as a result of this error (defective or incorrect end product)? Assume that the customer arrives home, discovers the incorrect error, and very unhappily returns to the restaurant.

Several costs are obvious, such as the cost of the incorrect food that cannot be resold (raw material, packaging, and end products), the employee's time to make new food (the correct order), and the manager's time to speak with the dissatisfied customer in some situations. Other costs are hidden and not fully apparent, such as the additional space and resources that the restaurant must keep on hand as extra inventory to handle these errors as they are basically throwing away food when a customer returns an order. Next there are hidden costs associated with preventing errors in the future; the manager may decide that another employee must double-check each order before it leaves through the drive-through. Now, extra work hours or a new hire is required to prevent errors.

Other costs have a much deeper impact and are not immediately visible to the fast-food restaurant at the time of the returned order and are related to future sales from the dissatisfied customer. The customer with the incorrect order may have lost faith in the restaurant and no longer choose to buy food there and buy from a competitor instead. Lee Resource Center, a marketing research company in South Carolina, United States, reports that for every customer who bothers to complain, 26 other customers will remain silent to the company [2]. This is concerning because on average for every one dissatisfied customer, there are potentially 26 more in the marketplace of whom the business is not aware. Also, Lee Resource Center reports that the average "wronged customer" will tell 8–16 people about it and 91% of unhappy customers will not willingly do business with the company again [2]. This is alarming because one dissatisfied customer can now have a negative impact 16 times greater by sharing their negative experience with others. These lost sales have a major negative impact on revenue and profitability and can directly be traced to poor quality.

This fast-food drive-through quality failure provides a concrete example of how quality can be "free" if the restaurant manager were to have proactively taken these costs and lost revenue associated with a wrong order and devoted it to creating a "system" that inherently produced 100% quality without inspecting every order before it left the drive-through. As discussed later in the book, this could be accomplished through developing standard operating procedure, poka-yoke, employee training, and enhancing communication. With the fast-food drive-through example as a frame of reference, the costs of poor quality can be separated into four categories:

1. Prevention
2. Appraisal
3. Internal failure
4. External failure

Prevention costs are related to planning and training. In a perfect world, most of the quality costs for an organization should fall into this category.

This category is proactive and focuses on creating, implementing, and delivering on a system that produces 100% quality. In the fast-food drive-through example, the manager would ensure that his or her employees are fully trained and have the tools to make 100% correct orders. *Appraisal costs* are most closely associated with inspection costs. For example, in the fast-food drive-through example inspection costs are related to the employee who must double-check each order before it leaves the drive-through. A company cannot "inspect-in" quality; this is an after-the-fact gatekeeper. If an incorrect error gets to this stage, it prevents the error from reaching the end customer for *internal failure* costs will be realized. The internal failure costs include the cost of the material and time to make the incorrect food and the reporting back to the manager.

The final category, which is also the worst-case scenario and most expensive, is an *external failure.* An external failure results when a defective or incorrect product reaches the end customer. In this case, the order must still be remade, but as discussed earlier, sales and profitability are now negatively affected. Figure 4.1 is a diagram displaying the inverse relationship between the cost of quality by category and the negative impact on customer satisfaction.

The old saying that "An ounce of prevention is worth a pound of cure" is very pertinent to quality. Top-performing companies place most resources in prevention costs related to training and creating a system that produces zero defects. Lower-tier companies tend to spend most of their "quality" resources on defects, rework, inspection, and errors related to external failure.

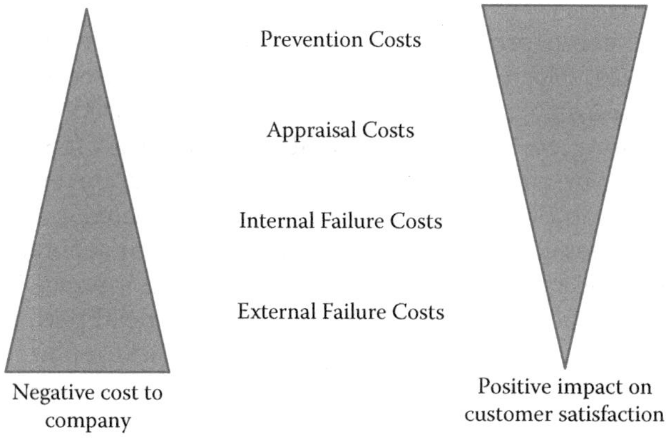

FIGURE 4.1
The costs of quality.

4.3 Cost Reduction and Customer Satisfaction Benefits of Lean

Lean is primarily a management approach to reduce operational costs through examining all forms of waste in an organization from the end customer's perspective. A key tool in implementing Lean manufacturing is the creation of a value stream map (VSM) that depicts the flow of every process from the customer's perspective. Any process that does not add value from a customer's perspective or is something that the customer is willing to pay for should be removed and every activity is categorized as "value-added" or "non-value-added." Back to the fast-food drive-through example, activities such as preparing the hamburger, placing it in the wrapper, and handing it to the customer add value from the customer's point of view. These are all necessary steps to deliver the hot sandwich to the customer. On the other hand, items such as double-checking the order before giving it to the customer (inspection) does not directly add value and this set could be eliminated if the process were created to eliminate errors and therefore inspection. Lean, by definition, focuses on the end customer and reducing costs that do not benefit the customer. In 2002, the US Army in coordination with Boeing implemented Lean manufacturing and saved an estimated $1.7 million per year that included a 45% reduction in cycle time, 28% reduction in labor costs, and an efficiency increase to 89% [3]. These actual savings from the implementation of Lean demonstrate the hard dollar value savings.

4.4 Profit Potential and Customer Satisfaction of Six Sigma

Significant increased profit potential exists for companies from the implementation of Lean Six Sigma. This is a result of increased customer satisfaction which translates into increased sales and increased profitability due to lower operating costs. Increased quality levels mean fewer dissatisfied customers, happier current customers, and lower scrap/report works. In 2010, the American Society for Quality (ASQ) conducted a survey of over 2,000 quality professionals to understand better the impact of implementing Six Sigma and the resulting improved business performance [4]. From the survey, 98% of respondents reported improved financial performance, 94% reported improved product/service quality, and 90% reported improved quality of customer performance [4].

4.5 Blending Lean and Six Sigma for Maximum Benefit

As mentioned at the start of this chapter, Lean Six Sigma blends the quality benefits of Six Sigma with the cost reduction benefits of Lean manufacturing. Any new program will require additional organizational resources in terms of money and time; the key idea behind Lean Six Sigma is that the recurring cost benefits outweigh the initial implementation costs. The cost to implement Lean Six Sigma can range dramatically based on an organization's approach to the implementation process. Some organizations choose to implement Lean Six Sigma as a pilot program on a limited scope basis, with a single project. The cost for such a pilot initiative can be very low, as typically only employee time and some process changes are needed. On the opposite end of the spectrum, full organizational implementation can require over $500,000 depending on the size of the organization due to new hires, consultants, and organization-wide training. Chapter 6 of this book describes deployment options in more detail; the goal for an organization is to implement Lean Six Sigma in the optimal way to reap the greatest return based on organizational constraints.

References

1. Crosby, P.B. (1979). *Quality Is Free: The Art of Making Quality Certain: How to Manage Quality—So That it Becomes a Source of Profit for Your Business.* New York: McGraw-Hill.
2. Lee Resource Center. (2014). *Customer Satisfaction Levels.* http://leeresources.com/
3. Jenkins, M. (2002). *Across the Enterprise Boeing Is Attacking Waste and Streamlining Process. The Goal? Cost Competitiveness.* http://www.boeing.com/news/frontiers/archive/2002/august/cover.html
4. Seibert, J.II and Schiemann, W.A. (2010). *Power to the People. Quality Progress.* http://asq.org/quality-progress/2010/04/customer-satisfaction-and-value/power-to-the-people.html

5

Metrics and Performance Measurement

5.1 Introduction

This chapter discusses metrics and performance measurement as they relate to the overall implementation of Lean Six Sigma programs. To successfully implement a Lean Six Sigma program, organizations must be able to quantify quality levels and costs to compare these amounts with similar industries. A common business adage summarizes this well, stating, "If you can't measure it, you can't manage it." Due to significant variation between industries, no single quality or cost standard exists that applies to every organization. For example, a jet engine manufacturer looks at a quality failure very differently than a fast-food restaurant. This variation can make quality improvement a very personal journey for an organization. Currently no comprehensive or universal quality measures exist for all companies. Various US government agencies at the state and local levels often use nonscientific and nonstandardized approaches to estimate quality. Metrics that track the effectiveness of the Lean Six Sigma program are a very important component in the process to ensure that goals are being met and that data-driven decisions are being made. The metrics that are created for the Lean Six Sigma program should be developed to provide valuable information on the success, variability, and efficiency of the process. The metrics should gauge whether initiatives are accurately measuring the success of the programs.

5.2 Six Sigma versus Three Sigma Quality Levels

A primary goal of Six Sigma is to reduce the variation of processes and products from the end customer's perspective. For example, with the US Postal Service, if you mail a letter from New York to Los Angeles, you should expect it to arrive in four business days (this is the average or mean delivery time). The letter will not always be delivered in four business days, there will be a range or variation in the actual delivery time, such as the letter being delivered in as few as two days or as many as six days (the range or standard deviation of the delivery time). One of the primary goals of Six Sigma is to reduce this variation to zero; in other words, the letter will be delivered in four business days every time. This will allow customers to better plan their business processes and their expectations of the US Postal Service. Relating this to a Sigma level, a Sigma quality level measures the variation or spread of a given process in terms of the standard deviation of the process and how well it meets customer expectations. The higher the Sigma level is, the better. So in the case of the US Postal Service, assume that the mean delivery time is four days ($\mu = 4$) and the standard deviation is plus or minus two days ($\sigma = 2$).

Assume that the customer needs the letter to be delivered within two to six days. We can calculate the US Postal Service's Sigma level, by dividing half the customer's requirement (delivered within two to six days) by the standard deviation of the process (two days). In this example, the US Postal Service is operating at a One Sigma level. If the Postal Service were able to reduce the standard deviation of their process from two days to one day ($\sigma = 1$), the US Postal Service would now be operating at the Two Sigma level. If the Postal Service could reduce variation even more, they could operate at a higher Sigma level.

Prior to the introduction and proliferation of Six Sigma in the 1990s, three Sigma quality levels were the acceptable benchmark for quality. In other words, a company could operate at the three-Sigma level and be considered a quality leader in the field. Now, a Six Sigma level is the quality expectation and "stretch" goal for many organizations. From a statistical perspective, a Six Sigma operating level corresponds to the number of defects generated, often times referred to as defects per million opportunities. If a company is operating at a Six Sigma level, the company will produce 3.4 defects per million parts produced by the company. In contrast to the Six Sigma operating level, the Three Sigma quality level translates into 2,700 defects per million parts produced. Considering the complexity of modern processes, Six Sigma quality isn't optional; it's required if the organization is to remain viable [1]. The requirement of extremely high quality is not limited to multiple-stage manufacturing processes; consider what Three Sigma quality would mean if applied to other processes [1]:

- Virtually no modern computer would function.
- 10,800,000 healthcare claims would be mishandled each year.
- 18,900 US savings bonds would be lost every month.
- 54,000 checks would be lost each night by a single large bank.
- 4,050 invoices would be sent out incorrectly each month by a modest-sized telecommunications company.
- 540,000 erroneous call details would be recorded each day from a regional telecommunications company.
- 270,000,000 (270 million) erroneous credit card transactions would be recorded each year in the United States.

To help maintain Six Sigma levels of quality several parameters are worthy of tracking to gauge an organization's effectiveness:

- Key process output variables
- Key process input variables
- Estimated completion date
- Actual completion date
- Tools utilized for the project
- Open action items
- Baseline measurements
- Financial calculations
- Actual project environment benefits and financial savings
- Lessons learned concerning project effectiveness

From this information, managers can gain insights into the effectiveness of each project and the overall success. Successful projects can also be achieved for use in later projects. In general, metrics need to capture the right activity and be designed to give immediate feedback. Successful metrics focus on the process rather than the product or individuals.

Reference

1. Pyzdek, T. (1999). *The Six Sigma Handbook.* New York: Quality.

6

Deployment Alternatives

6.1 Introduction

The economic and quality benefits of Lean Six Sigma are clear. Many organizations can significantly improve their bottom-line performance and improve customer satisfaction by focusing efforts using the Lean Six Sigma methodology. This chapter covers the next phase of that process, after the need has been identified that cost and quality improvement are important organization goals. The next phase is implementation and execution. There is a variety of alternatives to deploy Lean Six Sigma. These alternatives range from a massive organizationwide launch that covers multiple facilities to a smaller scope, short-term project. The following is a list of deployment alternatives ranging from large-scale organizational initiatives to smaller project-based launches.

- Corporate-wide launch
- Single-facility launch
- Department-based launch
- Single-product–based launch
- Project-based launch

Each method has advantages and disadvantages and the deployment mechanism should be selected based on the available resources (financial, employee, and technology), the project timeline, the project goals, and the corporate culture of the organization. Table 6.1 summarizes key benefits and drawbacks of each method.

For an organization just beginning the Lean Six Sigma process, the project-based launch is most often recommended. The primary reasons for this are that it requires fewer resources and has a shorter timeline. The key idea is that early quick results can lead to bigger projects in the future based on the success of smaller projects. A project is also easier to manage, for example, examining quality defects arriving from a single overseas supplier. Other advantages include:

TABLE 6.1

Deployment Method Benefits and Drawbacks

	Benefits	Drawbacks
Corporate-wide launch	1. Biggest results 2. Consistent processes across the entire organization 3. Optimizes the entire business system 4. Economy of scale	1. Highest cost 2. Longest implementation time 3. Requires intensive planning 4. Requires intensive data collection
Single facility launch	1. Can be used as a pilot project for other facilities 2. Analyzes entire facility for complete optimization	1. High cost 2. Requires intensive data collection
Department-based launch	1. Simplified implementation analysis 2. Promotes departmental team work	1. Can miss opportunities in other departments
Single-product–based	1. Analyzes more than just emissions from a facility	1. Data collection intensive
Project-based launch	1. Short timeline 2. Low cost	1. Smallest impact

- Utilizes tools in a more focused and productive way
- Increases communication between management and practitioners
- Facilitates a detailed understanding of critical business processes
- Gives employees and management views of how Lean Six Sigma tools can be of significant value to organizations

The central concept, regardless of which deployment method is chosen, is to tie project results into bottom line and quality benefits. The same concept is applied when prioritizing organizational needs or selecting among potential projects. These benefits should be expressed in terms of key process output variables such as:

- Process variation
- Process cycle time
- Cost per unit
- Customer satisfaction
- Scrap or defect rates

Several key concepts to keep in mind during the deployment planning phase include:

- Communicating the benefits of the project as a business strategy across the organization
- Aligning with management in the deployment of Lean Six Sigma
- Building a successful infrastructure for Lean Six Sigma deployment

- Selecting and implementing successful Lean Six Sigma projects and project teams
- Utilizing the right metric to drive the right activity
- Planning and execution of projects
- Selecting the right statistical tools

Finally, the recommended overall approach to the deployment of Lean Six Sigma is based on proven Six Sigma methodologies. The Six Sigma process involves the DMAIC methodology described below:

DEFINE

Define and clarify the project goal and timeline.

Establish a cross-functional team.

MEASURE

Define the current state and current processes, including the development of a process flow map to baseline the system and to identify any bottlenecks.

Collect and display data including task time, resources required, and process statistics.

ANALYZE

Determine process capability and speed utilizing statistical tools and charts.

Determine sources of variation and subsequent time bottlenecks.

Identify and quantify value-added and non-value-added activities.

IMPROVE

Generate ideas.

Conduct experiments and validate improved processes.

Develop action plans and standard operating procedures.

CONTROL

Develop control plan.

Monitor performance.

Mistake-proof processes.

This provides a brief introduction to the general approach for deploying Lean Six Sigma using Six Sigma. A detailed overview of the Lean Six Sigma implementation road map is discussed in Chapter 8.

6.2 Choosing a Lean Six Sigma Provider or Partner

When an organization chooses to implement Lean Six Sigma, an outside organization can be very useful and cost effective to help with the implementation. This may appear to be an expensive proposition when compared to an in-house implementation. However, the organization has to consider the real cost and time required for effective program development, trainer development, and so forth. In addition, it is important for organizations to consider the value of time lost when instructional material is not satisfactory, not effective, or is inefficiently presented.

A Lean Six Sigma consultant can help with determining the deployment strategy, conducting initial training, and providing project coaching. The decision of which group is chosen can dramatically affect the success of the program. However, choosing the best group to help an organization implement Lean Six Sigma can be a challenge. Often the sales pitch given by a consultant sounds good, however, the strategy or training do not match the needs of the organization.

The following is a suggested list of questions to present to the Lean Six Sigma consultants being considered:

- What is your basic Lean Six Sigma strategy and implementation plan?
- What do you suggest doing next if we would like to compare the Lean Six Sigma program offered by your organization with that offered by other organizations?
- What reference material do you utilize that follows the information used in your program (so that people can get further information or review on concept at a later date)?
- During Lean Six Sigma training do you use multimedia presentations such as Power Point or hands-on activities?
- What is the basic format of your Lean Six Sigma course for executive training?
- How do you address business and service processes?
- What topics do you cover in your workshops?
- How do you address the application of the techniques to real-world situations?
- What software do you use?
- What have others said about your training/consulting in the application of Lean Six Sigma? Can they be contacted?
- What is the experience level of your staff?
- What companies have you helped successfully implement Lean Six Sigma?

Once the list of providers is narrowed down, consider requesting that they describe their basic implementation strategy to prioritize projects within organizations. It is also recommended to visit each prospect to see firsthand a one-day training session. Send a decision maker to view this session. Some providers or consultants might initially look good, and then appear less desirable after their training approach and material are reviewed first hand.

6.3 Essential Elements of Deployment Plan

Applying Lean Six Sigma so that bottom-line benefits are significant and incur lasting change results requires a well thought out and detailed plan. Once a Lean Six Sigma consultant is selected (if used), he or she should assist the organization with the development of an effective implementation plan. Table 6.2 displays a sample implementation schedule for the kickoff plan. Although every plan is unique, it should consider the following essential elements, which are described in detail in subsequent chapters.

Several of the key elements for the Lean Six Sigma implementation plan are:

- *Create support infrastructure for the first wave:* The support infrastructure for Lean Six Sigma is the first and critical step to ensure a successful program launch. For Lean Six Sigma to be successful

TABLE 6.2

Kick-Off Schedule for Energy and Greenhouse Gas and Energy Minimization

Task	Timeline (days)
Choose a provider	30
Executive training	3
Project champion training	5
Create support infrastructure for first wave	21
Select Lean Six Sigma project(s)	21
Lean Six Sigma training: Measurement	5
Executive/champion meeting: Preanalysis phase	3
Lean Six Sigma training: Analysis phase	5
Executive/champion meeting: Preimprovement phase	3
Lean Six Sigma training: Improvement phase	5
Executive/champion meeting: Precontrol phase	3
Lean Six Sigma training: Control phase	5
Lean Six Sigma training/infrastructure post-mortem	14
Lean Six Sigma training wave	14
TOTAL	**137 days**

there must be a commitment from upper-level management and an infrastructure that supports that commitment. The key factors for creating a successful infrastructure are discussed in the next chapter.

- *Select key players:* Deployment of Lean Six Sigma is most effective with the assistance of trained practitioners in the field. This can be accomplished by hiring a consultant. Internally, the organization should select a project champion to serve as the team leader of a diversified cross-functional team that has knowledge in the various business areas of the organization, including operations, engineering, human resources, maintenance, finance, and accounting.
- *Select key projects:* Initial projects should focus on relevant product lines or services that have high visibility within the facility. This will help to build confidence within the team and strengthen support within the organization. Some good examples would be studying customer satisfaction levels for a high-selling product. Early small wins will lead to future big wins.
- *Training and coaching:* Prior to training registration, team members should have been assigned to a defined project on which they will be working. This will help them to understand specifically how to utilize tools as they apply to the process they are analyzing. It will also give them an opportunity to ask detailed questions and receive coaching on their projects.
- *Project report-outs:* Although difficult to schedule, ongoing top-down and bottom-up communications are also essential elements of a deployment plan. Communication plans are often the first thing overlooked on a busy manager's schedule, but are essential to break down the barriers between managers and practitioners.
- *Post-mortem:* Successful deployment of Lean Six Sigma is best achieved in a series of waves focusing on strategic change areas. Between waves there is time for evaluating effectiveness, compiling lessons learned, and integrating improvements into the infrastructure. Some examples of improvement opportunities are:
  - Were projects aligned with the strategic focus of the organization?
  - Was communication adequate between practitioners and management?
  - Were the forecasted timelines met?
  - Did the phase achieve the desired results?

The deployment phase planning process and implementation are critical steps to Lean Six Sigma implementation. This phase lays the foundation for all other steps, provides the needed training, and establishes communication channels. Poor planning in this phase can lead to entire program failure.

TABLE 6.3

Lean Six Sigma Calculator Comparison

Calculator Name	Funding	Market Served	Fees	Input Data	Output Data
iSixSigma Process Calculator	For-profit	Manufacturing, service, and nonprofit	No cost	Defects and opportunities	DPM and Sigma level
Statistical Solutions Six Sigma Calculator	For-profit	Manufacturing, service, and nonprofit	No cost	Defects and opportunities	DPM, short- and long-term Sigma levels
KnowWare QI Macro Calculator	For-profit	Manufacturing, service, and nonprofit	Free 30-day trial; fee thereafter	Defects, opportunities, confidence levels	DPM, short- and long-term Sigma levels

6.4 Six Sigma Calculators Available on the Internet

6.4.1 Overview of Calculators and Comparisons

Several websites have emerged to assist organizations in implementing Six Sigma. These websites typically utilize a calculator format that requires input from the user, performs calculations based on the entered data, and then provides output of the Sigma quality level. Several sites offer recommendations and case studies as well. Similarities and differences exist among the calculators and significant differences may exist in their output (Table 6.3). Major categories of differences and similarities among the calculators include:

- The calculator developer (government, nonprofit, or private company)
- Market served (home, commercial, manufacturing)
- Fee for usage
- Support provided
- Calculation method

The remaining sections provide a description of each calculator, including the website address, fees, and calculation methods.

6.4.2 iSixSigma Process Calculator (http://www.isixsigma.com/process-sigma-calculator/)

The iSixSigma is a for-profit company that was founded in 2000 in Ridgefield, Connecticut; it is a diversified media business that provides essential information, research, and how-to knowledge to help businesses and organizations worldwide improve execution [1]. The iSixSigma Sigma Process Calculator is a free website service that estimates the defects per million opportunities (DPMO), defect percentage, yield percentage, and process Sigma level based on opportunity and defect quantities entered by the user.

6.4.3 Statistical Solutions Six Sigma Calculator (http://www.statisticalsolutions.net/dpm_calc.php)

Statistical Solutions, LLC is a for-profit company that provides Six Sigma consulting services to organizations [2]. This free calculator is similar to the iSixSigma calculator, but generates additional information using more advanced statistical data. For example, the user enters the number of defects observed (n) and the total size of the sample or population (N) and the calculator estimates the defects per million (DPM), short-term process Sigma metric (σst), and the long-term process Sigma metric (σlt). The long-term process Sigma metric (σlt) includes the 1.5 Sigma shift.

6.4.4 KnowWare QI Macro Calculator (http://www.qimacros.com/lean-six-sigma-articles/sample-size/)

KnowWare is a for-profit process improvement consulting company founded in 1996 and based in Denver, Colorado [3]. Of all the available online calculators, this calculator offers the most advanced features and advanced statistical calculation capabilities in the form of an Excel spreadsheet using macros. KnowWare offers a free trial version, but the full version requires a fee. The calculator is able to use confidence levels, attribute data, and variable data; it includes integrated statistical process control (SPC) tools. The calculator can draw control charts, Pareto charts, histograms, fishbone diagrams, and flowcharts [4].

References

1. iSixSigma. (2014). *iSixSigma Process Calculator*. http://www.isixsigma.com/process-sigma-calculator/
2. Statistical Solutions, LCC. (2014). *Statistical Solutions Six Sigma Calculator*. http://www.statisticalsolutions.net/dpm_calc.php
3. KW Knowware. (2014). *KnowWare QI Macro Calculator*. http://www.qimacros.com/lean-six-sigma-articles/sample-size/
4. KW Knowware. (2014). *QI Macros*. http://www.qimacros.com/qi-macros/

7

Overview of Statistical Analysis

7.1 Introduction

Data-driven decision making is a key feature of all aspects of Lean Six Sigma. The DMAIC cycle relies heavily on statistical analysis and statistical tools to measure and improve business processes. This chapter provides an introduction and overview to common statistical concepts that are relevant during the implementation of Lean Six Sigma. An understanding of these concepts, such as mean, variance, and hypothesis testing are critical to the successful implementation of Lean Six Sigma and many of the Six Sigma "tools" imply an understanding of these concepts.

7.2 Discrete and Continuous Data

Discrete versus continuous data is a key concept in statistical quality control with regard to quality control charts. The "type" of data that one is collecting has a major impact on how the data are analyzed and the Six Sigma tools to be used. A simple method to distinguish between the two is that discrete data are always whole numbers, such as the number of red candies in a dish, the number of customer complaint calls received at a call center, or the number of defective headlights coming off an automotive assembly line. On the other hand, continuous data are not whole numbers, typically involve decimals, and can be measured to different levels of accuracy, such as someone's drive to work every day; the drive distance could be reported as 10 miles, 10.125 miles, or a very accurate 10.2554512 miles. Other examples of continuous data include part dimensions (such as the exterior dimension of a piston), the wait times (such as the time to get your food at a fast-food restaurant drive-through), or the temperature outside. Once the type of data (continuous versus discrete) has been determined, the appropriate methods to collect and statistically analyze the data can be established as discussed in the following sections of this chapter.

7.3 Means and Variance

Means and variances are critical calculations that apply to both discrete and continuous datasets. A mean is an average value of a sample, for example, the average outer diameter of a piston (continuous) or the average number of customers who arrive at a fast-food drive-through per hour (discrete). The variance is a measure of the spread or range of the data. It measures how close that the entire data are to the average or mean value. The variance provides insight into the magnitude of the spread and provides a comparison to means and ranges. For example, suppose the average household income in a particular state is $55,000 per year and the range of incomes is $10,000 to $250,000 (a very large range). Now, suppose that in the same state the average household income for entry-level engineers is also $55,000, but the range is $45,000 to $65,000 per year. The conclusion can be made that even though the averages for the two groups are the same, the range or variance are very different. It would be much easier for someone to guess the average household income for an entry-level engineer versus the average household income of an entire state because of the difference in the ranges. Variances provide this information; additionally, means and variance work together to provide a complete picture of the dataset in terms of the expected value (mean or average) and the range (variance) that the dataset covers.

The equation to calculate the mean ($\bar{x}$) is:

$$\bar{x} = \frac{\sum_{i=1}^{n} x_i}{n}$$

where

n = sample size.

x_i = each unit of the sample (first unit in the sample to the last unit).

The equation to calculate the variance is:

$$\sigma^2 = \frac{1}{n}\sum_{i=1}^{n}(x_i - \bar{x})^2$$

As mentioned, the variance (σ^2) provides information regarding the spread or range of the data. The standard deviation is the square root of the variance and it provides information on the "average" spread or dispersion within the dataset. The mean, variance, and standard deviation are best understood with an example. Suppose that you pack carrot sticks in your lunch every day over a five-day workweek, but that you do not bring the same number of carrot sticks each day. Table 7.1 provides the data regarding the number of carrot sticks that you pack every day.

TABLE 7.1
Example Dataset

Sample Number	Work Day	x_i	Number of Carrot Sticks Packed
1	Monday	x_1	7
2	Tuesday	x_2	5
3	Wednesday	x_3	4
4	Thursday	x_4	3
5	Friday	x_5	8

From the dataset, the calculation for the mean is:

$$\bar{x} = \frac{\sum_{i=1}^{n} x_i}{n} = \frac{(7+5+4+3+8)}{5} = 5.4$$

From the dataset, the calculation for the variance is:

$$\sigma^2 = \frac{1}{n}\sum_{i=1}^{n}(x_i - \bar{x})^2$$

$$= \frac{1}{5}\left[(7-5.4)^2 + (5-5.4)^2 + (4-5.4)^2 + (3-5.4)^2 + (8-5.4)^2\right] = 1.85$$

From the dataset, the calculation for the standard deviation is:

$$\sigma = \sqrt{\sigma^2} = \sqrt{1.85} = 1.36$$

Now, to place things in context of the example, you bring an average ($\bar{x}$) of 5.4 carrot sticks to work each day with an "average" range (σ) of 1.36 carrot sticks. Notice that the mean provides an expected value (5 or 6 carrot sticks per day) and the standard deviation provides an expected range per day (5 or 6 carrot sticks per day plus or minus 1 or 2 carrot sticks). Also notice from the variance and standard deviation equations that the minimum value for both is zero (variance and standard deviation cannot be negative). A variance and standard deviation of zero indicates that there is no spread or variation in the data; in other words, you bring 5 carrot sticks to work every day, no more or no less (every x_i is equal to 5). As a reminder, one of the primary purposes of Six Sigma is to reduce variation (σ) and to make processes standardized and predictable, so every day instead of bringing exactly 5 carrot sticks to work, the US Postal Service is delivering a letter from New York to California in exactly three business days every time a letter is mailed. The next sections of this chapter discuss other methods to measure and predict variation in data.

7.4 Random Variables and Probability Distributions

Random variables are variables that can assume a value based on chance or probability. The value is not fixed and can assume a range of values. For example, the time to be served at a fast-food drive-through, the number of days for a letter to be delivered, or the outside machined diameter of a piston. The range of values that these random variables can assume is described or predicted by probability distributions. A probability distribution is a mathematical formula that best predicts the mean and variance of a random variable. In other words, a probability distribution is a function that assigns probabilities to the values of a random variable. Random variables and probability distributions are critical concepts related to Six Sigma and statistical process control. The manager or engineer interested in improving a process must first understand and baseline the parameter that he or she is interested in improving, such as the number of days for a letter to be delivered from New York to California. The parameter (number of days for the letter to be delivered) is the random variable and the expected value and range (one to five days) is the probability distribution. There are several very common probability distributions used in Six Sigma for both discrete and continuous random variables. Tables 7.2 and 7.3 provide brief descriptions of the common probability distributions and their use in process control for discrete and continuous datasets, respectively.

TABLE 7.2

Common Discrete Probability Distributions

Distribution Name	Uses and Examples
Binomial	Uses: Estimating the probabilities of an outcome in any dataset for success or failure trials
	Example: The number of light bulbs that will fail in a shipment of 100 light bulbs
Geometric	Uses: Inspection sampling in large batches to find the first good item
	Example: The number of job candidates interviewed for an open position before finding the first good candidate
Negative binomial	Uses: Inspection sampling in large batches to find the first defective item
	Example: The number of good circuitboards inspected before finding the first defective circuitboard for laptop computer manufacturing line
Poisson	Uses: Reliability tests and failure tests
	Example: The expected number of defects that a automobile company will produce based on a historical average number of defects
Hypergeometric	Uses: Inspection sampling in small batches to find the first defective item
	Example: The number of defects in a shipment of 10 engines for final assembly at an automotive plant

TABLE 7.3

Common Continuous Probability Distributions

Distribution Name	Uses and Examples
Normal	Uses: Probability assessments between independent events and inferential statistics
	Example: The expected time and interval for a student or group of students to finish a midterm exam
Exponential	Uses: Mean time between failure, queueing theory, arrival time estimates
	Example: The expected number of customers who will arrive for service at a fast-food restaurant during lunchtime
Lognormal	Uses: Machine downtime estimation
	Example: The expected time that a drilling machine will be nonoperational in a facility for a given month
Weibull	Uses: Reliability and failure estimation
	Example: The comparison of mean failure times for two light bulb manufacturing companies based on mean failure times

Of these distributions, the most prevalent distribution is the normal distribution. Section 7.5 is devoted to describing its use in Lean Six Sigma and statistical process control.

7.5 Normal Distribution

The normal distribution, often referred to as the "bell curve" is the most common and most often used probability distribution in Lean Six Sigma. It is characterized by its bell-shaped curve as displayed in Figure 7.1.

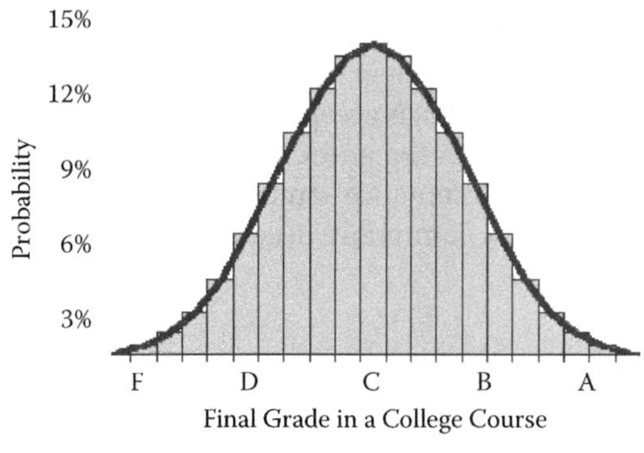

FIGURE 7.1
Normal distribution curve.

The normal distribution is an excellent predictor for many occurrences in nature, manufacturing, and service. The normal distribution curve is read from left to right and the height of the curve corresponds to the probability that the random variable will assume the value along the horizontal axis. In other words, the bell curve represents the chance that an event or value will occur. Notice that most events or values will occur at the average or middle of the curve and fewer occurrences will happen at the extreme left and right ends of the curve. To put this in context, think of a college class. Most students will have Cs (in the middle or top of the bell curve) and fewer students will receive As or Fs (the tail ends of the curve on the far left and right) as displayed in Figure 7.1.

The normal distribution is one of the most commonly occurring distributions in nature and in general processes. For example, the heights and weights of children, the diameter of a machined part, and the life of a light bulb are all predicted by the normal distribution. Due to its ability to predict accurately many occurring events in nature and process management, it is heavily utilized in Lean Six Sigma to compare means, predict future results, validate improvements, and to test hypotheses.

7.6 Sampling and Sample Sizes

As mentioned earlier, Six Sigma relies heavily on data analysis to draw conclusions and to improve processes. Oftentimes, one of the first steps in this process is to determine how much data to collect in order to evaluate and validate conclusions statistically. The goal is to determine the most effective sample size that will minimize data collection time and costs while ensuring that the results are statistically significant. Sample size equations are introduced in Chapter 8 as part of the Lean Six Sigma methodology and rely on the normal distribution presented earlier. Sample sizes have a critical role in acceptance sampling. Acceptance sampling is a reactive quality control process that involves randomly selecting several items in a large lot to judge whether the entire lot meets required quality standards. If the few randomly selected items meet the quality standard the entire shipment (lot) is accepted; if the few randomly selected items do not meet quality standards the entire shipment (lot) is rejected, sent back, and payment is not made. Acceptance sampling is an application of the normal probability distribution and commonly used in the automotive industry.

7.7 Confidence and Prediction Intervals

Like acceptance sampling, confidence and prediction intervals rely on the normal probability distribution to infer data regarding an entire population from a sample or subset of the population. Confidence intervals provide the expected range of an "average" parameter for a population at a specific confidence level. For example, a manufacturer of pistons could be 95% confident that the average machined length for all the pistons that her plant produces will be between 75.1 cm to 75.3 cm. Prediction intervals are similar, but instead of estimating the average value for the entire population, prediction intervals provide the expected range for one given sample from the population at a specific confidence level. For example, the piston manufacturer could be 95% confident that the next piston produced on her manufacturing line will be between 74.9 cm to 75.5 cm. Prediction intervals (for a single unit) are always larger than confidence intervals (the average of the population). Both confidence and prediction intervals have a key role estimating conformance to customer requirements and provide a solid estimate of how well processes will perform with respect to the customer requirements. It is a great tool to predict the number of quality failures and cost of poor quality. Mathematics and additional concepts are introduced in Chapter 8.

7.8 Hypothesis Testing

Hypothesis testing is used in Lean Six Sigma for screening potential causes and to compare and evaluate process improvements statistically. A hypothesis test calculates the probability, p, that an observed difference between two or more data samples can be explained by random chance alone, as opposed to any fundamental difference between the underlying populations from which the samples came [2]. Hypothesis testing is used to infer statistical conclusions at a given confidence interval. For example, a toothpaste manufacturer could use hypothesis testing to make the claim, "Our toothpaste results in 25% fewer cavities as compared to competitors." Chapter 8 discusses hypothesis testing in more detail.

7.9 Design of Experiments

Design of experiments (DOE) is a branch of applied statistics that deals with planning, conducting, analyzing, and interpreting controlled tests to evaluate the factors that control the value of a parameter or group of parameters [3]. DOE helps to identify the key parameters that affect a desired outcome. For example, what parameters have the largest influence when baking a nice-looking and great-tasting cake? DOE could be used to analyze different combinations of parameters (ingredients, oven temperature, mixing time, bake time, etc.) to determine the best combination to get the best cake. DOE is heavily used in Lean Six Sigma to identify, isolate, and control optimal parameters and is discussed later in Chapter 8.

References

1. iSixSigma. *iSixSigma Process Calculator*. (2014). http://www.isixsigma.com/process-sigma-calculator/. http://www.isixsigma.com/process-sigma-calculator/. Retrieved February 22, 2014.
2. DMAIC Tools. (2014). *Hypothesis Testing*. http://www.dmaictools.com/dmaic-analyze/hypothesis-testing. Retrieved on February 22, 2014.
3. American Society for Quality. (2014). *Design of Experiment*. http://asq.org/learn-about-quality/data-collection-analysis-tools/overview/design-of-experiments.html. Retrieved on February 22, 2014.

8

General Approach to Implement Lean Six Sigma

8.1 Introduction

Lean Six Sigma is a comprehensive and flexible system for achieving, sustaining, and maximizing business success [1]. It is uniquely driven by a close understanding of customer needs, disciplined use of facts, data, and statistical analysis, and diligent attention to managing, improving, and reinventing business processes [1]. Six Sigma strives to improve quality, productivity, and bottom-line financial performance. The Six Sigma approach provides a methodology to achieve these successes and provides a system to base these improvement initiatives for corporations, nonprofit groups, and government agencies. Six Sigma relies heavily on data, facts, and the use of statistical tools to study whether an improvement has been made. These statistical tools are a powerful aid in conducting experiments and comparing data and providing important information about a process to find the causes of problems and draw conclusions. Lean Six Sigma methodologies are broken into six fundamental categories:

- Define products or services.
- Know the stakeholders and customers and their critical needs.
- Identify processes, methods, and systems to meet stakeholders' critical needs.
- Establish a process of doing work consistently.
- Error-proof process and eliminate waste.
- Measure and analyze performance.

To achieve these fundamental goals, Lean Six Sigma uses the five-step DMAIC methodology as described in Chapter 6.

Lean Six Sigma has three overall targeted solutions: process improvement, process design/redesign, and process management. Process improvement refers to a strategy of fixing a problem while leaving the basic structure of the process intact [1]. Process design/redesign is a strategy not to fix a process but to replace the process to fix the problem [1]. Process management refers to integrating Lean Six Sigma methods into everyday business including:

- Processes are documented and managed "end-to-end" and responsibility has been assigned in such a way as to ensure cross-functional management of critical processes.
- Customer requirements are clearly defined and regularly updated.
- Measures of outputs, process activities, and inputs are thorough and meaningful.
- Managers and associates use the measures and process knowledge to assess performance in real-time and take action to address problems and opportunities.
- Process improvement and process design/redesign are used to raise the company's levels of performance, competitiveness, and profitability constantly.

The remainder of this chapter discusses the general Six Sigma approach from the systems' perspective. The chapter relates the Six Sigma approach to the DMAIC cycle.

8.2 Overview of Lean Six Sigma Approach and Systems

After the need has been identified to implement Lean Six Sigma and top management supports the initiative the implementation can begin. Baseline data collection is one of the most important steps of the Lean Six Sigma process because the data generated provide the team and management with a much greater understanding of the key problems and areas of opportunity. Before rushing in to collect baseline data, proper planning should be done to ensure the scope, goal, and timeline of the project fit into the strategic plan of the organization. This chapter provides the systems approach framework to ensure that these goals are met using the five-step Six Sigma DMAIC process discussed earlier. The remainder of this chapter expands upon the systems approach and breaks this framework into smaller pieces, providing an execution model and process flow to implement Lean Six Sigma. As a reminder, the systems approach examines the organization as a whole, or a sum of all business processes to achieve established goals. The concept is to use data to develop comprehensive system-wide changes that will drive operational, quality, and economic performance versus routine incremental improvements. In addition, several examples are provided to further explore and explain each step of the framework. The Lean Six Sigma process consists of 11 steps:

1. Establish Lean Six Sigma team and charter.
2. Review existing records and data related to the charter goals.
3. Create process flowcharts and conduct throughput analyses.
4. Collect process data within the facility related to the charter goals.
5. Analyze the data by work unit or area to establish baseline data.
6. Identify major quality and cost improvement opportunities.
7. Determine, evaluate, and select process, equipment, and method improvement alternatives for implementation.
8. Develop the Lean Six Sigma deployment and execution plan.
9. Execute and implement the Lean Six Sigma plan and timeline.
10. Validate the program versus goals.
11. Monitor and continually improve performance.

Figure 8.1 separates this 11-step process as it relates to the Six Sigma approach.

By applying Six Sigma and the systems approach, meaningful and relevant alternatives can be developed and evaluated in a standardized manner to improve quality, enhance customer satisfaction, reduce costs, and improve financial performance. With this approach, alternatives are fully described, improvement impacts are quantified, a feasibility analysis is conducted, a financial justification analysis is performed, and feedback is collected from all stakeholders before making a final decision to implement. To aid in describing the 11-step process, a case study is utilized and an example given to discuss the real-world application of each step. The case study involves a company in the Midwest that is a leading manufacturer of industrial floor-cleaning equipment. Approximately 30,000 units are produced per year in a plant that was opened in 1955 and employs 85 people.

8.3 Step-by-Step Implementation Guide

This section provides a step-by-step guide to implement Lean Six Sigma by applying the 11-step framework discussed earlier. The following subsections describe each step in more detail including a discussion of how to implement relevant Lean Six Sigma tools and mathematical analyses in the context of the case study.

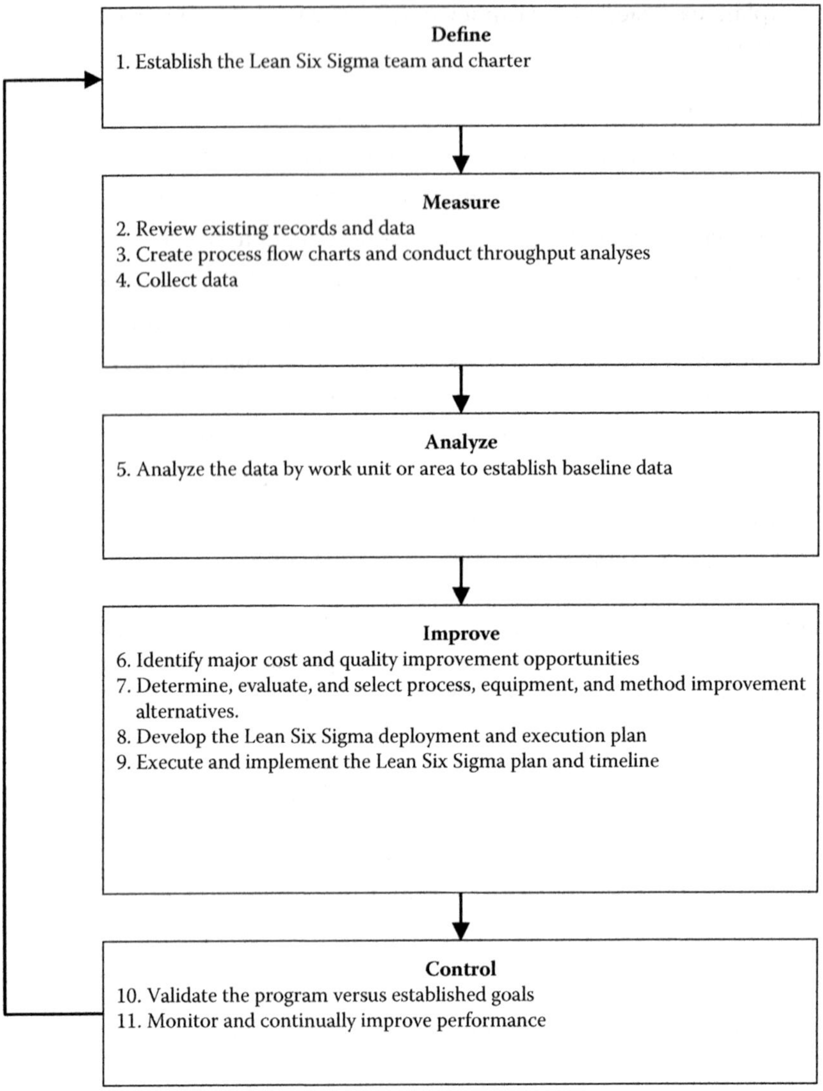

FIGURE 8.1
Eleven-step Lean Six Sigma approach.

8.3.1 Step 1: Establish Team and Define Project

The primary purposes of the step are related to the DMAIC Define stage and involve gathering sufficient information to clarify the opportunity for improvement, learning about the process, and learning about the organizational barriers to solving the problem. It also involves developing the initial plan and framework to address the problem and lays the foundation for all future steps. After identifying the need to minimize solid waste and gaining top management support, the first step in the process involves establishing the team and defining goals. The key outcomes and deliverables of this step are:

- A letter of support from top management to all employees
- Team leader and member identification
- Initial training for the team
- Problem clarification and goal identification with metrics
- Team charter
- Project timeline
- Project budget

8.3.1.1 Upper Management Support

Upper management support is one of the most important components for any successful project. A letter, memo, or e-mail from the organization president, CEO, or facility manager to all employees is a great way to set the tone of the project and to create awareness among all employees. A clear message must be sent that Lean Six Sigma, customer satisfaction, and cost competitiveness comprise an important focus of the company and that all employees are requested to take part and contribute. Figure 8.2 shows a sample letter that may be used.

8.3.1.2 Project Team

Selecting the team to lead the Lean Six Sigma project should not be made lightly. Thought should be put into selecting a cross-functional team that has process knowledge and good interpersonal skills. The team leader should be in a management position at the facility. This will allow for faster communication with management and a higher degree of authority within the team and throughout the facility. Efforts should also be made not to select an overburdened manager; the team leader must have adequate time that he or she can devote to the project and lead all meetings.

The team itself should be comprised of five to six core members representing different areas in the facility, including both management and hourly employees. For example, in a manufacturing plant, a janitorial worker, an engineer, maintenance technician, a production supervisor/manager, a line employee, and an accountant would be a good mix. The team members should exhibit concern for customer satisfaction, possess strong interpersonal skills, and be well respected among their peers. The team will serve as the internal cheerleaders for the project to help build excitement, participation, and awareness. In addition, a technical expert within the company or

January 1, 2015

Memo to all facility employees

Dear coworkers:

As president of the organization, I am pleased to state that we have made a strong, unified commitment to collaboration and cooperation in improving our quality and enhancing profitability. This commitment has been demonstrated in several projects, such as our Six Sigma green belt certification program and the hiring of management and staff to support related quality areas.

In light of the recent successes and focus on customer satisfaction, we have placed increased emphasis on unifying our quality and productivity efforts throughout the organization. As a first step to accomplish the goal, we are forming a Lean Six Sigma team comprised of your peers to evaluate our current performance and implement process changes to increase our quality, reduce operating costs, and reinforce our focus on customer satisfaction. This project will allow us to move forward with these key initiatives and any input that you have is highly welcomed and appreciated. If you have any input, please contact your immediate supervisor to share your thoughts.

In conclusion, we have provided strong support for quality and cost enhancements. We are very pleased with the success of our collaborative efforts at our headquarters office and intend to implement a strong program at our facility. Our continued commitment to the environment will sustain the efforts of this program so that we may become a true quality and cost leader in our field.

Sincerely,

President and CEO

FIGURE 8.2
Sample letter.

contracted as a consultant is a value-added addition to the team and highly recommended. Temporary team members may be added to the project to assist in data collection during the Measure phases discussed later in the chapter. For the example case study, the team was led by an engineering manager and included a consultant hired on an "as needed basis," a line worker, a custodian, an accountant, a process engineer, the safety captain, and a production supervisor. This cross-functional team collectively possessed a strong understanding of all core and support operations within the facility.

8.3.1.3 Initial Training and Introductory Meeting

The team will require initial training before beginning the project. The goals should be clearly outlined, including the metrics, timeline, and budget. The team should also have the opportunity to give input to the team charter and help in its development. The purposes of the training and introductory meeting are to:

- Introduce the team and exchange contact information/work schedules.
- Establish clear SMART goals (simple, measurable, agreed to, reasonable, and time based).
- Define initial team member roles.
- Finalize the budget.
- Finalize the timeline.
- Create the team charter.

The technical expert should take the lead role in this process to ensure an adequate training and a clear understanding of the project goal and the path ahead. The training itself should focus details on the 11-step process presented earlier in Figure 8.1.

8.3.1.4 Problem Clarification and Goal Identification with Metrics

The goals of the project should be clearly expressed as soon as possible. This will provide the team with much needed direction and serve as the gauge to evaluate all team activities and accomplishments. The goals should provide a specific direction for the project and not vague generalized improvement slogans such as "To become a quality leader in the automobile engine manufacturing field" or "To reduce the organization's operating costs." The goal should be a more specific, SMART goal, such as:

- Reduce the quality defects in the paint finishing unit by 5% per year from the baseline year of 2014.
- Increase productivity by 15% in all final assembly areas from the baseline year of 2014.

The measure phase is to identify correct measures, establish a baseline, and eliminate trivial variables [2]. The following tools are important to understand:

- Basic statistics
- Statistical thinking
- Cost of quality

- Measurement system analysis
- Critical parameters
- Critical to quality

There are two types of statistics, descriptive and inferential. The descriptive statistics summarize the historical data [2]. Basic descriptive statistical analysis consists of the mean, median, mode, range, variance, and standard deviation [2]. Also, measuring the cost of quality is critical because high variations and inconsistencies can cause high cost and waste valuable resources. Inferential statistics are based on analysis of the sample to infer performance of the process [2]. It is generally concerned with the source of the data and seeks to make generalizations beyond the data at hand. Inferential statistics can include regression analysis and hypothesis testing among many others. Many common tools are used in this step that involve interacting and listening to internal and external customers to frame the problem or issue to be solved properly. Some of the common tools used in this process include:

- Kano model
- Affinity diagrams
- Pareto analysis
- Surveys
- Force field analysis

A Kano model is a tool developed to compare the relationship between customer satisfaction and customer requirements [2]. The requirements are separated into three groups related to basic features, performance features, and "excitement" features that differentiate the product or service in the marketplace. The Kano model is read left to right and depicts the time and resource requirements to transition from a basic product or service to an exciting, market-leading, and innovative product or service. Figure 8.3 provides a sample of a Kano model.

The affinity diagram is a method used to obtain input from stakeholders related to a product or service. It is a tool that allows all team members to have an equal voice related to brainstorming ideas for process or product improvement. Using this method, a group can generate ideas very quickly and group them based on themes to identify action quickly for creating solutions [2]. Figure 8.4 displays an affinity diagram example. The diagram is read from left to right and each box represents a different theme and potential action items.

Pareto analysis is a tool used to make decisions based on possible impact and importance to customer satisfaction and bottom-line results. It is known as the 80:20 rule which means that usually 80% of the problems come from 20% of the processes. The Pareto chart is a bar chart showing attributes of

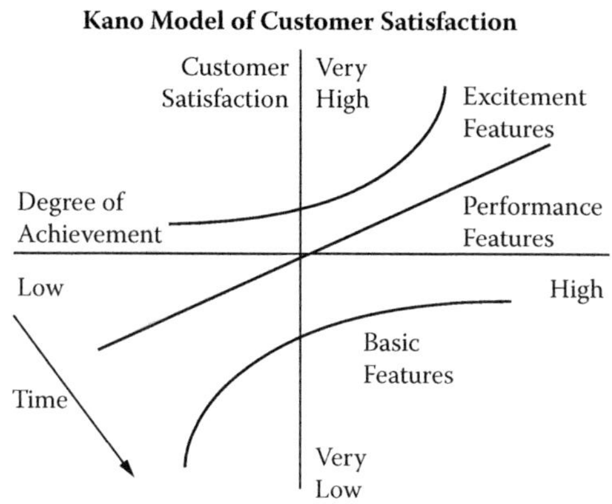

FIGURE 8.3
Kano model.

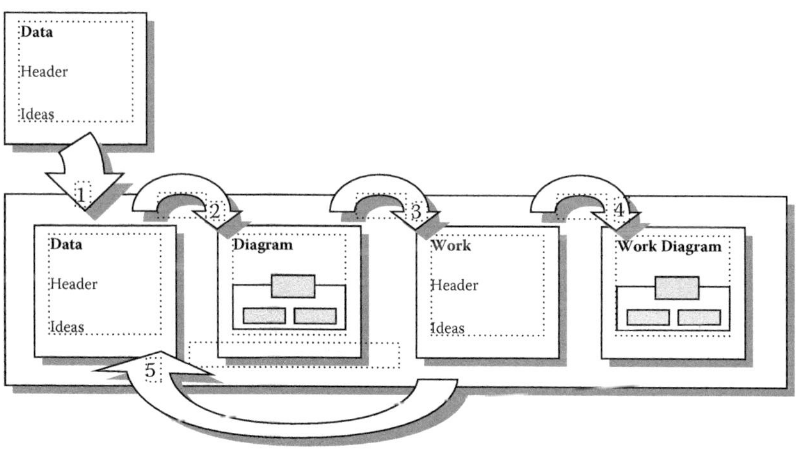

FIGURE 8.4
Affinity diagram example.

the problem on the *x*-axis and frequency of occurrence on the *y*-axis [2]. Figure 8.5 provides an example of a Pareto analysis; it lists the attributes, such as causes of defects or product sales from left to right, with the highest occurrences listed on the left side. Pareto analysis is a great tool to discover the primary causes of defect quickly for a Lean Six Sigma team to focus on in order to maximize results.

Figure 8.6 displays a force field analysis. The force field analysis is a method to identify supportive and resistive resources that could be effectively utilized toward the process goals [2]. The objective is to identify factors to accelerate change or resistive resources that would slow process change.

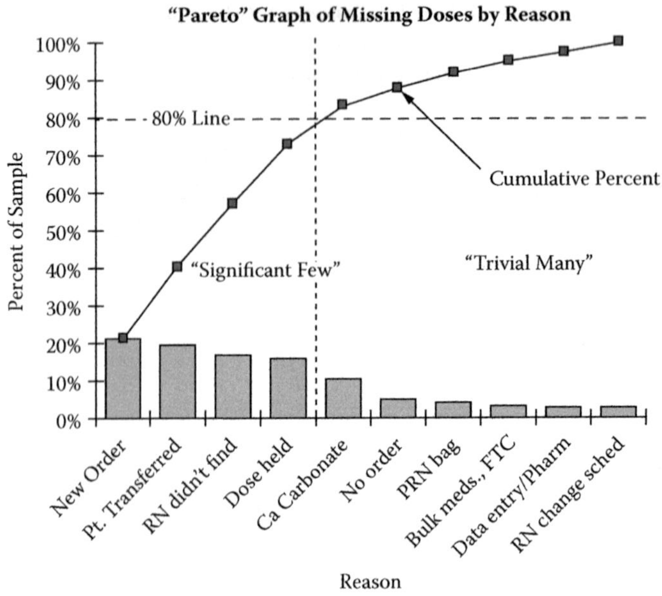

FIGURE 8.5
Pareto analysis.

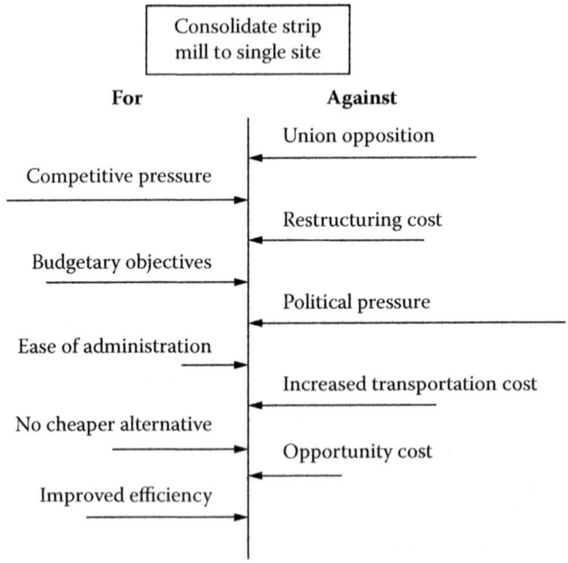

FIGURE 8.6
Force field analysis.

It is similar to the SIPOC* and process map, but focuses more on factors that positively or negatively influence change. It is read like a tug-of-war game with the forces on the left promoting and justifying the need to change and the forces on the right resisting or hindering change. This force field analysis is a great tool for executives to view the process from a very high level and to identify the best methods to promote Lean Six Sigma with employees and stakeholders.

For the case study project, the goal was established to reduce defects in the "Champ" line by 8% over a 12-month period versus the same period last year without increasing operating costs. Such a clear goal provides a specific goal for the team and the constraints to achieve it. In this case, the metrics were defects per unit and dollars.

8.3.1.5 Team Charter

The team charter is a statement of the scope, objectives, and participants in a project. It provides a preliminary description of roles and responsibilities, outlines the project objectives, identifies the main stakeholders, and defines the authority of the project manager. It serves as a reference of authority for the future of the project. The purpose of the team and project charter is to document:

- Reasons for undertaking the project
- Objectives and constraints of the project
- Directions concerning the solution
- Identities of the main stakeholders

The three main uses of the project charter are:

- *To authorize the project:* Using a comparable format, projects can be ranked and authorized by return on investment.
- *To serve as the primary sales document for the project:* It provides stakeholders with a one- to two-page summary to distribute, present, and keep handy when fending off other projects that may attempt to consume the allocated resources.
- *As a focus point throughout the project:* The charter may be used as an introductory document for new team members, provide a focal point during team meetings, and be used in control or review meetings to ensure tight scope management.

In addition, a project charter compiles the problem definition, goals, objectives, and action plans to achieve them. It is a road map that:

* Suppliers, Inputs, Process, Output, Customers

- Justifies the project efforts with financial impact
- Describes the problem and its scope to be addressed by the project in the specified timeframe
- Declares the goal, objectives, and measures of success
- Defines the roles of the team members
- Establishes the timeline, milestones, and key deliverables
- Identifies required critical resources

For the example company, the team charter can be seen in Figure 8.7.

8.3.1.6 Project Timeline

The project timeline is one of the key deliverables from the Define stage. The timeline tracks project performance versus established goals and serves as the strategic implementation plan. The timeline should be viewed as a control document to evaluate the progress of the team versus pre-established milestones. Proper planning is needed to ensure the timeline is achievable and will meet the goals of the project. Below are some general guidelines for the time required for each step in the Lean Six Sigma implementation process. Please note that the timeline assumes that the team will be devoting approximately 30% of their time to the project and 70% of their time to normal job duties. Based on the size or scope of the project, completion times may vary from a few months to several years. In the example displayed in Table 8.1, the timeline required 35 weeks for the case study.

TABLE 8.1

Project Timeline

Lean Six Sigma Step	Time (weeks)
1. Establish the Lean Six Sigma team and charter.	2
2. Review existing records.	2
3. Create process flowcharts and conduct throughput analyses.	2
4. Collect additional facility data.	2
5. Analyze the data by work unit and establish baseline.	3
6. Identify improvement opportunities.	6
7. Evaluate and select alternatives.	6
8. Develop the deployment and execution plan.	2
12. Execute and implement the Lean Six Sigma plan.	8
13. Validate the process versus goals.	1
14. Monitor and continually improve performance.	1
TOTAL	**35**

Lean Six Sigma Team Charter

Purpose Statement and Team Objectives:

The mission of the team is to reduce quality defects in the Champ Line by 8% versus 2010 levels without added expenses. The team is committed to working effectively together by monitoring process effectiveness, following through on commitments, and helping one another learn. The mission will be accomplished with the following objectives:

- To complete the 11 step assessment process for the company by March 31.
- To prepare first draft proposals, and present to CEO by April 15.
- To refine proposals, and present to regional management meeting on April 25.
- To present the final plan to the CEO by May 15.

Team Composition and Roles

The team will be made up of representatives from each functional area in the plant including management and craft employees. This range of skills and knowledge will enable the team to understand the issues relating to individual work units, as well as developing solutions to the problems outstanding.

Jack Smith will take the role of team leader. In that role he is responsible for:

- Ensuring this Team Charter is abided by;
- Managing the day to day operations of the team and the team's deliverables;
- Managing the budget;
- Providing support and assistance to individual team members; and
- Providing status reports to the CEO on a weekly basis.

Authority and Empowerment

Jack, as team leader, has the authority to direct and control the team's work, and team members are allocated full time to this project, for its duration.

Resources and Support Available

A budget of $4,000 is available to cover travel and initial supplies to launch the program. The CEO will meet with Jack Smith at 4:30 pm every Monday afternoon for a progress update and to provide support and coaching appropriately.

Operations

- The first team meeting will be on Monday, February 28 at 2:00 pm.
- The team will meet every Monday afternoon from 2:00 pm to 3:30 pm for the duration of the project.
- Each member is expected to present a short status report for the aspect of the project they are working on.
- If a member is unable to attend, a notification must be sent to the team leader and someone else designated to report on the status and communicate further expectations.
- A summary of each meeting will be prepared by Jack and e-mailed to all members by the morning following the meeting.

FIGURE 8.7
Sample team charter.

8.3.1.7 Project Budget

The final deliverable from the Define stage of the process is the budget. The budget identifies the financial resources available to the project for process improvements and data collection assistance. The budget could include such items as:

- Training and reference materials
- Outside training
- Data collection tools and quality monitoring tools (stopwatches, clipboards)
- Data collection labor
- Equipment purchases
- Communication funds (newsletters, banners, posters)
- Travel funds to visit top-performing sites for benchmarking
- Meeting refreshments (if desired)

Several approaches exist when creating the budget. For example, project start-up funds could be allocated by management and spent at the team's discretion. Management could allocate $1,000 to $10,000 for the items listed above and empower the team to best utilize the funds. Many successful companies have taken the approach that no additional dollars will be allocated to the project. For the case study example, the team was given $2,000 by executive management to cover tools and program startup costs.

8.3.2 Step 2: Review Existing Records and Data

Reviewing the existing organizational records for quality, productivity, and cost usually provides significant insight into company performance and baseline data. Very useful data and information can be gained to help focus the efforts of the team and eliminate the need for the collection of existing raw data, which can be very time consuming and expensive to collect. Collecting high-quality records will save a great deal of time, money, and effort versus raw data collection. The types of records to collect include:

- Purchasing, inventory, maintenance, and operating logs
- Supply, equipment, and raw material invoices
- Equipment service contracts
- Repair invoices
- Major equipment list
- Production schedule (representative of a year)

- Company brochure or product information
- Material safety data sheets (MSDS)
- Facility layout (hardcopy plus CAD copy, if available)
- Process flow diagrams
- Quality records
- Scrap and defect rates (including quantities and costs)

The purposes of reviewing these records are to calculate baseline data and to determine additional data collection needs related to the Lean Six Sigma project goal. The records to be reviewed will vary based on the project goal. For example, if the Lean Six Sigma project goal is to reduce quality defects by 8% in a specific manufacturing line, then the team would focus on collecting records pertaining to scrap rates, production rates, machine utilization, worker performance, and customer quality reports. A review of purchasing records can also be beneficial to backtrack into an estimate of production rates if detailed information is not available. The team should be aware of several limitations that exist from the existing records, such as:

- Cost and quality data may not be separated by product; it may be generalized by work unit or facility.
- The material invoices and service records may not provide adequate data if the contracted company does not collect this information.

Figures 8.8 through 8.11 serve as the data collection worksheets for the existing record review. Figure 8.8 provides the recommended list of documents to collect during this step.

8.3.3 Step 3: Create Process Flowcharts and Conduct Throughput Analyses

The goal of this step is to aid the Lean Six Sigma team in fully understanding the business processes and capabilities of the facility. An understanding of these processes is crucial in developing alternatives to improve quality and reduce costs. A process flowchart is a hierarchical method for displaying processes that illustrates how a product or transaction is processed. It is a visual representation of the work flow either within a process or an image of the entire operation. Process mapping comprises a stream of activities that transforms a well-defined input or set of inputs into a predefined set of outputs.

A well-developed process flowchart or map should allow people unfamiliar with the process to understand the interaction of causes during the work flow and contain additional information relating to process improvement

Lean Six Sigma
Records Review

In order for the Lean Six Sigma Team to begin the assessment, please compile the following items and complete the attached questionnaire. If additional space is needed than what is provided on the questionnaire, please attach sheets.

Please provide the following:

1. Major Equipment List
2. Production Schedule (representative of a year)
3. Company Brochure or Product Information
4. MSDS Sheets
5. Facility Layout (hard copy plus CAD copy, if available)
6. Process Flow Diagrams
7. Receipts from each supplier and utility provider (representative of one year)
8. Please indicate which safety items the waste assessment team will need to enter your facility:
9. ☐ Safety Glasses
   ☐ Steel toe Shoes
   ☐ Hard Hats
   ☐ Hearing Protection
   ☐ Other ____________________

FIGURE 8.8
Existing records review worksheet—page 1.

(such as bottlenecks and points of failure). Process mapping is a way to identify the various activities of the process and show their interrelationships [2]. A well-defined map must address the following:

- Process has a purpose.
- Process has beginning and end states.
- Process has needs or inputs.
- Process must have a clear target performance.
- Process output does vary due to uncontrolled sources of variation.
- Process must be evaluated based on its mean or typical performance, as well as range between worse and better performance levels.

Figure 8.12 displays an example of a process map. Notice that it starts with the customer at the top and relates all processes from a value-added perspective as perceived by the customer. The process map is a great tool for the Lean Six Sigma team to understand how a product is made or a service is performed. It allows for a common frame of reference for the team to make decisions, identify improvement opportunities, and bottlenecks.

Section 1: Company Information

Date: ____________

1. Company Name: ____________
   Address: ____________
   Company Contact Name: ____________
   Phone: ____________ Fax: ____________
   Purchasing department contact: ____________ phone: ____________
   Janitorial services contact: ____________ phone: ____________
   Engineering department contact: ____________ phone: ____________
   Environmental services contact: ____________ phone: ____________
2. Primary SIC Code: ____________ Secondary SIC Code (s): ____________
3. Number of Shifts: ____________ Shift times: From: To:
   Employees Per Shift: First: ____________ First: ______ - ______
   Second: ____________ Second: ______ - ______
   Third: ____________ Third: ______ - ______
   Total Employees: ____________
   Days per year the plant is in production? ____________
4. What is the labor and benefit rate for hourly employees? (if necessary, break out classifications)
   ____
5. Square Footage of Facility: ____________
   What is the cost per square foot of production space or warehousing space? ____________
6. What is the number of products produced last year? (breakdown by product) ____________
   ____
   What is the company's Minimum Attractive Rate of Return (MARR) and tax rate? ____________
   ____
8. What are the approximate annual sales for your company? ____________
9. Has a Lean or Six Sigma assessment ever been conducted before? ☐ Yes ☐ No

Section 2: Forklift Data

1. Number of forklifts ____________
2. Fork lift fuel type: gasoline/propane/electricity/natural gas

FIGURE 8.9
Existing records review worksheet—page 2.

To create a process map, it is best to first define the starting and ending points of the process that will be studied (the process map will begin and end at these points). Oftentimes this may include working with suppliers and internal customers if the process to be studied encompasses multiple facilities. Figure 8.13 displays a supplier, input, process, output, and customer (SIPOC) diagram and is used to expand upon the process map to identify stakeholders in the operation [2]. The benefit of this tool is to identify all variables that affect the process's performance and then prioritize them so that the team may take action. The SIPOC diagram is a tool that allows for a

Section 3: Process Information

1. Briefly list your company's products and/or services: ____________________

2. How would you rate current levels of productivity and cost reduction?

☐ Average
☐ Above Average
☐ Below Average

3. Please explain any reasons for this variation in production or production trends: ________

4. Please indicate all areas which are present at this location:

☐	Storage	☐	Research and Development
☐	Design	☐	Shipping/Receiving
☐	Lab Work	☐	Manufacturing
☐	Office	☐	Fabrication
☐	Retail	☐	Inventory/Warehousing
		☐	Other ____________________

5. Please indicate all manufacturing processes currently at this location:

☐	Anodizing	☐	Grinding	☐	Pickling
☐	Coating	☐	Heat Treating	☐	Polishing
☐	Blending	☐	Kiln Firing	☐	Printing
☐	Brazing	☐	Machining	☐	Rolling
☐	Cleaning	☐	Milling	☐	Shearing
☐	Degreasing	☐	Mining	☐	Slitting
☐	Electroplating	☐	Molding	☐	Stamping
☐	Etching	☐	Painting	☐	Welding
☐	Extruding	☐	Paint Strip	☐	Other ____________

Section 4: Energy Prices and Usage

1. Electricity cost per kilowatt-hour________ and usage last year (kilowatt hours) __________

2. Piped Natural Gas cost per therm or 100 cubic feet _____ and usage last year ___________

3. Liquid Propane Gas (LPG) cost per gallon _________ and usage last year ____________

4. Fuel Oil cost per gallon _________ and usage last year ________________

FIGURE 8.10
Existing records review worksheet—page 3.

common frame of reference for the entire supply chain to better understand the process and the interconnections . The SIPOC diagram is typically created early in the Define stage as it "frames" the problem for the entire team to baseline improvement initiatives.

Once the team has determined the beginning and ending activity steps, the mapping process can begin. Oftentimes, interviews with operators and

Section 5: Corporate Considerations		
1. Does your company have a written quality policy?	☐ Yes	☐ No
If yes, does this policy include ISO 9001 integration?	☐ Yes	☐ No
2. Is there a cost reduction program in place?	☐ Yes	☐ No
Is there a quality management program in place?	☐ Yes	☐ No
Is there an employee involvement program to improve operations?	☐ Yes	☐ No

If yes, please explain your employee involvement program: ________________

3. What are the priority areas where quality and cost reduction efforts should be focused? ____

4. Please list any Six Sigma green belts, black belts, or master black belts in the organization:

5. Please indicate any barriers to the Lean Six Sigma implementation

- ☐ Inadequate performance
- ☐ Corporate policy
- ☐ Engineering specifications
- ☐ Price
- ☐ Other ________________

COMMENTS:

FIGURE 8.11
Existing records review worksheet—page 4.

front-line employees are required to understand the process steps and linkages better. Some key points to process mapping include:

- Keep it simple.
- Start at a high level first.
- Involve the people closest to the process.
- Walk through the process yourself.
- Think end to end.
- Work with a small group of three to seven people. A larger group can make the activity cumbersome.

Below are the three general steps to create a process map:

1. Begin by stating the intention or desired goal to create a process map for activities completed between start-step A and stop-step B.
2. Document the activities that people do between steps A and B on self-adhesive notes. Attach the notes to a wall, flipchart, or whiteboard in

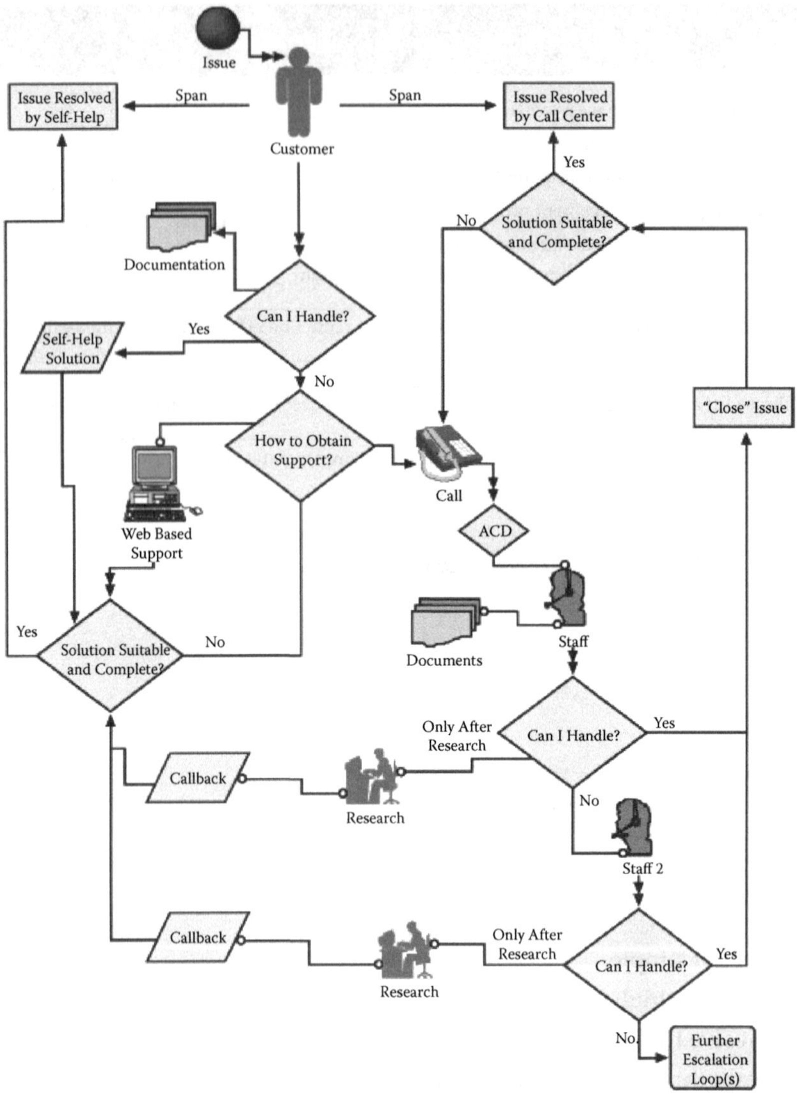

FIGURE 8.12
Process map.

the order in which they are normally completed. Begin documentation at a high level and then move into additional process maps that provide greater detail.

3. Document the process maps. Schedule a meeting to review the information. Make sure to verify and clarify the activities and their owners. Also, look for any immediate opportunities to create "quick wins."

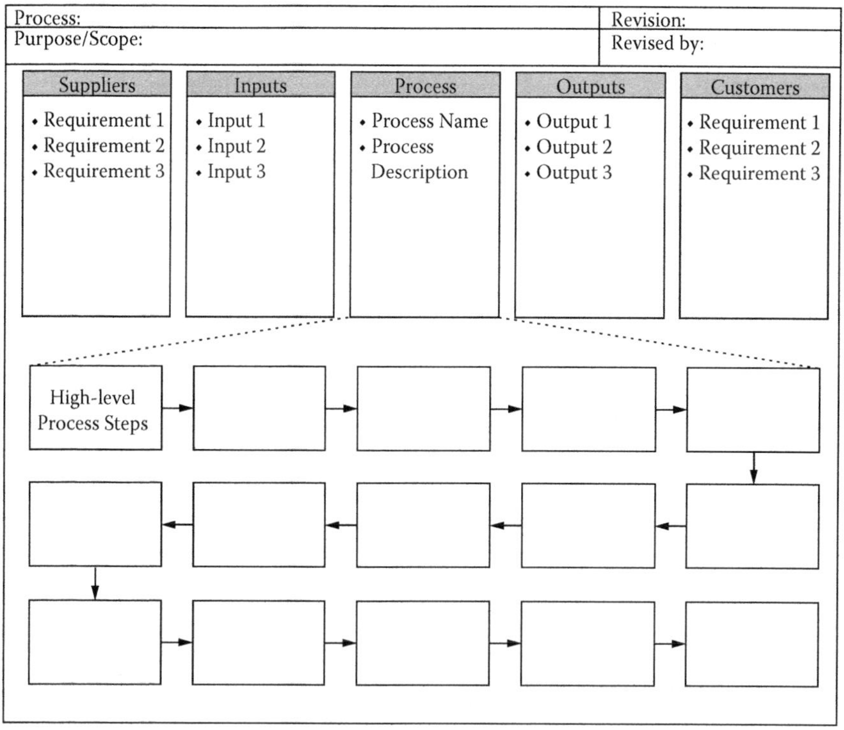

FIGURE 8.13
SIPOC diagram.

In order to identify areas for improvement, processes must be broken into subprocesses. A subprocess operates at a more detailed level than a core process and gets into who does what and why on a daily basis. Core processes must be broken down into enough detail to understand, monitor, manage, and analyze performance. As a general rule, processes must be described at a third level, as shown above, before improvement teams will be able to deal with them adequately. Figure 8.14 displays a data collection form that can be used when creating the process map and flowchart and Figure 8.15 displays the process charts that were created during the case study example.

Oftentimes when creating a process map, additional data need to be gathered to describe the process in terms of process times, production rates, and capacities. This may involve shop floor data collection, time studies, and throughput analyses. Section 8.3.4 provides additional details to collect data effectively and efficiently.

8.3.4 Step 4: Collect Process and Quality Data

On-site data collection is often necessary to complete the productivity and quality baseline for the facility. Depending on the amount of data available

Process Description:
Company/Location:
Recorded By:

Date:

Summary		
Category	No.	Dist.
Operations		
Transports		
Inspections		
Delays		
Storages		
Totals		

Step	Operation	Transfer	Inspection	Delay	Storage	Description of Method	Distance	Cycle Time
	○	⇨	□	D	▽			
	○	⇨	□	D	▽			
	○	⇨	□	D	▽			
	○	⇨	□	D	▽			
	○	⇨	□	D	▽			
	○	⇨	□	D	▽			
	○	⇨	□	D	▽			
	○	⇨	□	D	▽			
	○	⇨	□	D	▽			
	○	⇨	□	D	▽			
	○	⇨	□	D	▽			
	○	⇨	□	D	▽			
	○	⇨	□	D	▽			
	○	⇨	□	D	▽			
	○	⇨	□	D	▽			
	○	⇨	□	D	▽			
	○	⇨	□	D	▽			
	○	⇨	□	D	▽			
	○	⇨	□	D	▽			
	○	⇨	□	D	▽			
	○	⇨	□	D	▽			

FIGURE 8.14
Process flowchart form.

from previous records, on-site data collection needs vary. Some organizations keep very detailed records by work unit for productivity, quality, and cost. Via the data collected, the organization's productivity, quality, and cost profiles can be characterized, including annual levels and the percentage that each component contributes to the whole. These data are invaluable when evaluating cost-effective methods or process changes to drive improvement initiatives. This remainder of this section discusses the process to conduct on-site data collection, including the required preparation, tools required, a step-by-step guide, and also provides data collection forms.

8.3.4.1 Preparation for On-Site Data Collection

Before scheduling the data collection, some preparation work is required, including a team meeting to discuss several items to ensure a smooth project.

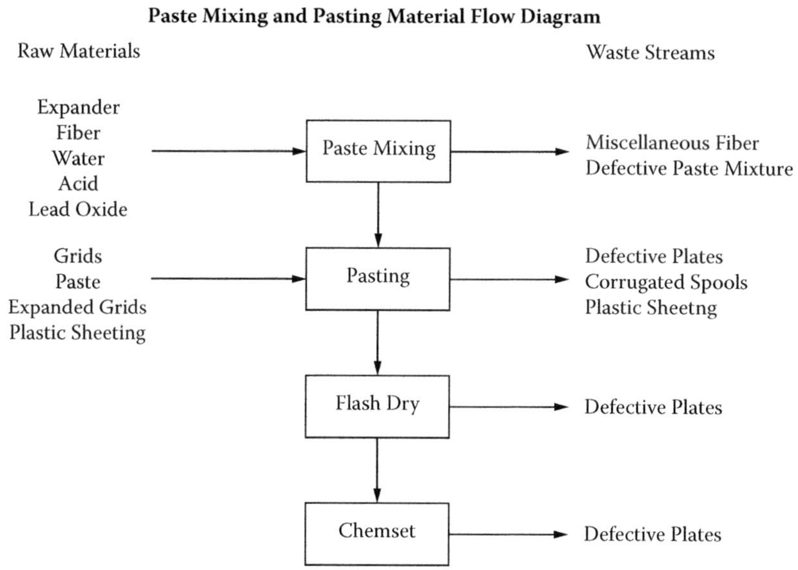

FIGURE 8.15
Process map example.

Specifically, the following items should be addressed and finalized prior to the data collection session:

- Hold a team meeting.
- Create an assessment plan and assign roles and facility areas of responsibility.
- Conduct training.
- Review the layout of the facility.
- Gather the needed tools and equipment.
- Assign data collection roles to temporary team members.
- Prepare the data collection forms.
- Hold a custodial staff meeting.
- Disseminate employee messaging regarding the assessment and expectations.

The team meeting prior to the assessment should be held to establish common goals and timelines. The primary outcomes from this meeting are training and development of the assessment plan. The training should be conducted by the technical expert and focus on the use of tools and the data collection form. The assessment plan should assign team members to the various areas of the facility to conduct data collection effectively and efficiently. The tools and equipment required during the data collection phase may include:

- Gloves
- Yardsticks
- Stopwatch
- Scales
- Clipboards (to hold the assessment forms)

Based on the size of the facility, it may be necessary to recruit additional support to collect data during the assessment. If additional help is used, these individuals should receive the same training as the core team. An approach that seems to work well is to assign one temporary team member from each work unit within the facility. For example, assign a supervisor or shop-floor worker from each production work area such as the metal stamping unit, the paint shop, and the accounting offices (work units will differ by business type). The advantages of assigning temporary team members within each work unit include faster and more accurate data collection. The temporary team member, as a member of the work unit in which data are being collected, will possess specialized knowledge regarding the productivity and quality issues for the specific work units.

The data collection form is one of the most important documents of the assessment process. The data collected with the form will be used to extrapolate the baseline data to determine the most significant areas for quality and cost improvement for the facility so care should be taken when collecting the data to ensure accuracy. At a minimum, the required information on the data collection form is:

- The date the data were collected
- The team members collecting the data (this is very useful if follow up or clarification is needed)
- Work unit and location
- Square footage of work area
- Number of employees assigned to the work area and shift days/times
- Equipment data (size, year built, energy consumption, usage per day)
- Notes and comments that may be useful when analyzing the data (including names and contact information for work unit members with specialized knowledge)

Figure 8.16 is a template of the data collection form that may be used during the data collection process.

A meeting should be held with the work unit supervisor to clarify the scope and support required for the assessment. On the day of the assessment,

Lean Six Sigma			
Data Collection Form			
Unit Name		Date of Study	
Location		Performed by	
Days of Operation			
Hours of Operation		Work Unit Supervisor	
Number of Employees		E-mail	
Square Footage of Work Area		Phone Number	
Production Rate (Units per Hour)			
Cycle Time (minutes)			
Machine Utilization			
Equipment Uptime			
Man Hours per Unit			
Defect Rate			
Rework Rate			
Recycled Materials			
Amount of Materials Recycled per Day			
Existing Energy and Emission Reduction Items:			
Notes:			

FIGURE 8.16
Data collection form.

the employees should be notified regarding the timing and data collection specifics. Notes and messages should be placed in the employees' work and break areas the day before and the day of the assessment to ensure compliance. During the meeting, the employees should be made aware of the goal of the project and the purpose of the assessment. Oftentimes, the front-line

March 24

Memo to All Employees:

On March 31 the Lean Six Sigma team will conduct an assessment at our facility between 2 p.m. and 5 p.m. The purposes of this assessment are to determine production and quality rates to help us become a stronger and more competitive organization. On the day of the assessment, please accommodate the team and provide assistance as requested. Also, if you have any ideas or suggestions to help us become more quality and cost focused, please share them with the assessment team on March 31.

Thank you for your support.

President and CEO

FIGURE 8.17
Sample service talk.

employees will have value-added suggestions and comments that will improve the study such as areas to target, bottlenecks, and common issues.

Finally, several days before the assessment, messaging in the form of e-mails, service talks, and postings in common areas should be disseminated to all facility employees discussing the details of the data collection process. The messaging should include the date of the assessment, and the purpose and process of the assessment. This will help ensure a smooth assessment by informing employees to maintain regular work practices and to prevent data collection delays resulting from the team continually being asked who they are and what they are doing. In addition, the messaging could solicit suggestions for improvement and employee comments. By sending the messaging earlier, it will give the employees time to think about Lean Six Sigma opportunities in their work units and provide more value-added feedback. Figure 8.17 is a sample service talk that could be used to inform all employees.

8.3.4.2 Assessment Guide

The assessment itself is straightforward: every work unit in the facility is analyzed and a data collection form is completed. During the walkthrough of the assessment, the team will gather raw data regarding their observations related to productivity, quality, and cost. The team also will examine the company's main production, the flow within the facility, number of product lines in the facility, and draw a process and material flow diagram. Ideally, if a layout of the facility is available it may be used to segment the plant to collect data for the team members. In general, a team of two people is recommended for each miniteam. This will allow one person to measure and the other to record the data.

8.3.4.3 Productivity Analysis and Time Studies

The primary purpose of the production analysis is to determine annual production rates (or capacities) and related cyclical patterns. Oftentimes, time studies are required to collect these data. Time studies involve timing the operation for one cycle of a product or service to estimate value-added and non-value-added activities. A time study shows how an employee spends his or her time working on different tasks. The analyst records a tally every minute in a specified area and observes the employee for a specified time. After the allotted time the analyst then records each task and can show the percentage of time the employee spends on each task. Work studies provide quantified process-level data to understand and baseline processes for improvement better. The information derived from the work studies flows into several other Lean Six Sigma analyses and aids in identifying bottlenecks.

The production rates provide a baseline related to productivity by each cycle, whether the cycle is a week, month, or quarter. Figure 8.18 is a capacity model worksheet that could be used to aid in this process. To calculate the capacity, the following information will be needed:

- Annual forecast
- Cycle time
- Efficiency (versus cycle time)
- Time available
- Utilization (uptime of the machine)

8.3.5 Step 5: Analyze Data by Work Unit and Establish Baseline

The primary outcomes of this phase are baseline data profiles for cost, quality, and productivity related to the work unit, product, or process to be improved. In a sense, establishing the baseline data is similar to getting a doctor's health checkup for a process or operation as it identifies parameters and metrics for the current state of the process. Just as the doctor may identify high cholesterol levels or blood pressure issues, this phase of the Lean Six Sigma methodology identifies areas where the process is "sick" and may need to be improved. These baseline data should be segmented by work unit of generation, process, or product. The key questions that are answered from this analysis are:

- What is the current cycle time for the product or process?
- What percentage of activities are value-added?
- What is the cost and profit to produce one unit?
- What is the capacity of the process?

Company Name
Title (Capacity Model)
Location (City, State)

Date:

Production Parameters	
Day per Year	233
Hours per Shift	8

Production Area		Time Required						Time Available			
Item #	Operation	Annual Forecast (Units)	Daily Forecast (Units)	Cycle Time (minutes)	Parts per Cycle (pcs/cycle)	Efficiency (percent)	Hrs Reqd per Day (hrs/day)	Shifts per Day (shift/day)	Utilization (percent)	Hrs Avail per Day (hrs/day)	Loading (percent)
1.0	Lathe	100,000	429	1.10	2	90%	4.37	2	90%	14.40	30.36%
2.0											
3.0											
4.0											
5.0											
6.0											
7.0											
8.0											
9.0											
10.0											
11.0											

FIGURE 8.18
Production analysis example.

- What are the machine utilization and uptimes related to the process?
- What is the quality acceptance rate for the process?
- What is the capability in terms of Sigma level?
- What types of process controls are used to improve process efficiency?

To answer these questions and to generate the baseline data, the existing records collected, the data gathered during the collection phase, and team member knowledge will be used. Additional data collection or verification may be required during the analysis portion. The first step in the data analysis process is to summarize the collected data in one document.

The most efficient manner to accomplish this is with the use of a linked spreadsheet. The data collected from the waste audit and records review is input into a linked spreadsheet to extrapolate the needed information. The information entered into this spreadsheet includes the data collected from each collection form during the collection phase. After the data are entered, the spreadsheet can be created to calculate annual values automatically based upon predetermined density factors. Sections 8.3.5.2 to 8.3.5.13 provide example calculations to establish the baseline in the context of the industrial floor-cleaning equipment manufacturer case study.

8.3.5.1 Example Overview

This case study applied the Lean Six Sigma methodology to redesign a final assembly work unit to increase capacity and reduce costs in a light manufacturing company located in the United States. The work unit that was analyzed for this case study was responsible for assembling an industrial-sized ride-on floor scrubber. The final assembly work unit that was redesigned for this study was responsible for 38% of the company's revenue stream and experienced various cost and process efficiency issues. These issues included an assembly line imbalance, floating bottleneck operations, safety concerns, high percentages of non-value-added time for employees, housekeeping issues, and storage issues. In addition, the company was seeking to increase production in this work unit to attain a larger market share. To accomplish these goals, the organization applied the Lean Six Sigma approach to create a meaningful and practical road map. The organization in this case study is a leading manufacturer of industrial cleaning equipment headquartered in the midwestern United States. The company began operations in the early 1900s and conducts business in more than 60 countries with over $3.1 million in annual sales and employs 155 individuals. Their operations are based on one shift, and its hourly workers are unionized. The company has explored options to relocate its operations to less expensive regions, such as Mexico or the southern United States, where they could save money by reducing labor costs and overhead due to increased costs and economic downturns. Before moving forward with these plans, the company investigated reducing costs

FIGURE 8.19
Current state layout for the case study.

and its location. Figure 8.19 provides a display of the current state layout for the process.

8.3.5.2 Cycle Time Analysis

Cycle time is the total time from the start of a process to the end of a process for one unit (product or service) to be produced, as defined by the Lean Six Sigma team or customer. The cycle time includes value-added process time, during which the product is being acted upon to bring it closer to an output, and non-value-added delay time, during which a product of work is waiting or in transportation. In other words, cycle time is the total elapsed time to move one unit of work (or product) from the beginning to the end of a physical process and includes both valued-added and non-value-added activities [3].

Cycle time is typically provided as an average value that includes a mean and standard deviation. It is typically determined via a time study, where several cycles are timed from beginning to end using a stopwatch. In general, a minimum of 10 to 20 cycles are required to determine an accurate baseline cycle time. Once the data are collected, the mean and standard deviation for one cycle are calculated. These values serve as the initial baseline or "current state" of the process that will be used to determine improvement possibilities.

For the case study, the cycle time to assemble one industrial floor scrubber was 229.5 minutes with a standard deviation of 27.0 minutes. The cycle began when the work unit operator retrieved the first subassembly (a faceplate) to begin assembly of the floor scrubber and ended when the operator tested the unit and stored it in the final goods storage area. Table 8.2 displays the cycle time calculation and time study for six cycles.

TABLE 8.2

Cycle Time Calculation and Time Study for Six Cycles

	Observations							
Element	**1**	**2**	**3**	**4**	**5**	**6**	**Mean**	**St. Dev.**
Get face plate	8	3	4	5	3	6	4.8	1.9
Get decal and (7691221)	15	5	5	2	6	5	6.3	4.5
Place decal	40	31	30	30	32	31	32.3	3.8
Place (7691221)	18	12	8	18	17	17	15.0	4.1
Tighten	8	6	4	7	7	6	6.3	1.4
Get (6491261) and (4890791) (×2)	7	6	6	6	6	6	6.2	0.4
Place (6491261) and tighten	24	24	26	24	25	25	24.7	0.8
Get screws	21	12	10	18	13	11	14.2	4.4
Get bolts and the red cable	11	10	12	11	11	11	11.0	0.6
Prepare (4890791) (×2)	18	21	20	19	20	19	19.5	1.0
Place (4890791) (×2)	26	11	23	14	17	21	18.7	5.7
Tighten	7	18	7	8	7	12	9.8	4.4
Add the red cable	9	8	7	7	7	8	7.7	0.8
Checking directions	16	45	21	12	35	36	27.5	13.0
Mistake	15	26	0	0	21	9	11.8	10.8
Store	16	10	13	15	13	15	13.7	2.2
Total	259	248	196	196	240	238	229.5	27.0

8.3.5.3 Machine and Labor Utilization Analysis

Machine and labor utilization rates are the percentage of time that equipment and workers are performing work or involved in activity. It is calculated by dividing the process time by the total time available to perform a process. For example, assume an assembly-line operator is working an eight-hour shift (480 minutes) and during that time he is responsible for assembling the final floor scrubber in the case study. Assume each assembly process requires 29 minutes and he assembles 12 units on the shift (328 minutes of assemble time). His utilization percentage is 72.5% (328 minutes/480 minutes). The same holds true for a machine or equipment. Assume in the same final assembly process that the operator uses a punch press that requires two minutes for each floor scrubber. For the 12 units produced, the processing time is 24 minutes. The utilization for the punch press machine is 5.0% (24 minutes/480 minutes).

Typically, the higher the utilization percentage, the better for the organization as it indicates that the resources (machines or workers) are involved in productive work that is generating a positive economic return. The utilization rates also provide insights into resource allocations and line balancing. If a worker or machine has a low utilization rate, the worker or machine may be assigned to another value-added activity on a different process. Conversely, if a resource has a very high utilization rate as compared to other resources, it may indicate that the resource is overloaded and potentially a bottleneck. The utilization percentage works closely with overall equipment effectiveness and capacity that are discussed in the following sections.

8.3.5.4 Machine Overall Equipment Effectiveness (OEE)

Overall equipment effectiveness (OEE) is a framework for measuring the efficiency and effectiveness of an automated or semiautomated process, by breaking it down into three constituent components: availability, performance, and quality [4]. OEE measures the utilization of a machine, but also combines machine availability (also known as uptime) and quality output of the machine into one metric. Following is the calculation for OEE based on three components:

1. Availability takes into account downtime loss, and is calculated as:

$$\text{Availability} = \text{Operating Time}/\text{Planned Production Time}$$

2. Performance takes into account speed loss, and is calculated as:

$$\text{Performance} = \text{Ideal Cycle Time}/(\text{Operating Time}/\text{Total Pieces})$$

Ideal cycle time is the minimum cycle time that your process can be expected to achieve in optimal circumstances. It is sometimes called design cycle time, theoretical cycle time, or nameplate capacity. Because run rate is the reciprocal of cycle time, performance can also be calculated as:

$$\text{Performance} = (\text{Total Pieces}/\text{Operating Time})/\text{Ideal Run Rate}$$

Performance is capped at 100%, to ensure that if an error is made in specifying the ideal cycle time or ideal run rate the effect on OEE will be limited.

3. Quality takes into account quality loss, and is calculated as:

$$\text{Quality} = \text{Good Pieces}/\text{Total Pieces}$$

OEE takes into account all three OEE Factors, and is calculated as:

$$\text{OEE} = \text{Availability} \times \text{Performance} \times \text{Quality}$$

OEE provides a percentage value from 0% to 100% measuring the overall strength of the machine in terms of adding value for the organization by taking into consideration machine uptime, performance to cycle time, and quality output. OEE is a great tool to identify weak or low-performing machines and pinpointing the cause of the low performance. For example, if a production run is one 8-hour shift (480 minutes), the machine experienced

45 minutes of downtime during the shift, the ideal run state is 45 pieces per minute, and a total of 15,000 pieces were produced (of that total 200 were rejected due to poor quality), the OEE is calculated as:

Availability = (480 minutes – 45 downtime minutes)/480 minutes = 90.63%

Performance = (15,000 pieces/435 minutes)/45 pieces per minute = 76.63%

Quality = (15,000 pieces – 200 pieces)/15,000 pieces = 98.67%

OEE = 90.63 × 76.63 × 98.67 = 68.53%

From the example, the performance of the machine is very low (76.63%) compared to the other OEE metrics indicating that the machine is not able to achieve the optimal cycle time. This serves as a benchmark for the Lean Six Sigma team and a first area for the team to target to improve the overall performance of the machine. OEE is also an excellent tool to compare the performance of a group of machines to identify weak links or areas for improvement in the facility.

8.3.5.5 Throughput, Capacity, and Demand Analysis

Throughput and capacity analysis determine the maximum output of a process or facility over a given timeframe. It is based on the number of machines, workers, shifts, utilization, and efficiency. Capacity analysis is oftentimes compared to demand and forecasted demand to determine if the process is capable of meeting current or future demand based on labor and machine constraints. A simple calculation for capacity is (number of machines or workers) × (cycle time) × (number of shifts) × (utilization) × (efficiency).

A more detailed and useful calculation for machine capacity examines equipment usage versus capability. The amount of equipment required for a process is referred to as the equipment fraction. The equipment fraction can be determined for a process by dividing the total time required to perform the process by the time available to complete the process. The total time required to complete the process is the product of the standard time for the process and the number of times the operation/process is to be performed [6]. Standard time is calculated by using time studies and the equation for the equation fraction is displayed below.

$$F = S * Q * \frac{1}{H * E * R}$$

where

F = number of machines required per shift to perform the task.

S = Standard time (minutes) per unit produced.

Q – Number of units to be produced per shift.

H = Amount of time available per machine.

E = Actual performance, expressed as a percentage of standard time.
R = Reliability of machine, expressed as percent "uptime."

The capacity and machine fraction should then be compared to current demand and forecasted demand to ensure that the process is capable of meeting demand. If there is a very large difference in capacity versus demand action may need to be taken to increase capacity if demand is too high or decrease capacity and cost if demand is too low versus capacity.

8.3.5.6 Value-Added Process Analysis

Lean manufacturing is a systematic approach to identifying and eliminating waste, which is also known as non-value-added activities through proper continuous improvement techniques [6]. The three factors that were evaluated for work to be considered as value-added were [7]:

- *Capacity:* Machinery, resources, tools, and employees used in the process must have the necessary capacity to produce a final product adding value.
- *Information/Instructions:* Employees must know their final product and process to achieve the final product with minimum waste or non-value-added activities.
- *Materials:* The material given to workers should be free from defects and able to be processed and sold on the commodity market. A worker should know which raw material is acceptable or unacceptable.

The eight forms of wastes that are considered non-value-added include [8]:

1. T—Transport: Moving people, products, and information
2. I—Inventory: Storing parts, pieces, documentation ahead of requirements
3. M—Motion: Bending, turning, reaching, lifting
4. W—Waiting: For parts, information, instructions, equipment
5. O—Overproduction: Making more than is immediately required
6. O—Overprocessing: Tighter tolerances or higher grade materials than are necessary
7. D—Defects: Rework, scrap, incorrect documentation
8. S—Skills: Underutilizing capabilities, delegating tasks with inadequate training

The value-added ratio measures the percentage of value-added activities versus the total activities for a process. The value-added ratio is the primary metric used in Lean manufacturing as it provides the basis to maximize value from all processes from the end customer's perspective by minimizing waste in every process and maximizing value.

8.3.5.7 Activity-Based Costing Analysis

Activity-based costing (ABC) is an accounting methodology that assigns costs and revenues to activities based on the actual consumption of each activity. The primary goal of ABC is to assign indirect costs, known as overhead, into direct costs compared to conventional costing. For example, if an automobile parts supplier produces four different parts: pistons, brake shoes, crankshaft pulleys, and exhaust valves, a manager, accountant, or engineer using conventional costing would examine three different costs: (1) direct costs for materials per part, (2) direct costs for labor per part, and (3) overhead. Using ABC, the company could dig deeper into the cost structure to separate the overhead in a more meaningful way by part type. So, in the example, the company could apply ABC to determine that the costs per part are as follows:

Labor costs:	\$2, \$3, \$5, and \$9, respectively
Materials:	\$5, \$3, \$7, and \$8, respectively
Maintenance (for machine):	\$5, \$2, \$6, and \$10, respectively
Engineering:	\$8, \$11, \$4, and \$13, respectively

The company has now used ABC to separate overhead into maintenance costs and engineering costs per part (these are usually not specified in a general overhead value). The sum of these four costs is the manufacturing cost per part. Overhead will still exist (for such items as the cost of insurance, lighting, rent on building, etc.), but can be directly related to the manufacturing cost (e.g., 20% of the manufacturing cost). The sum of the manufacturing cost and the overhead cost is the total cost of making the part. So if the company wants their profit to be equal to 30% of the total cost, they could base the profit requirements on the ABC model. The sum of the total cost and the profit is the selling price of the part. In essence, ABC allows managers, engineers, and accountants to better understand the complete cost and revenue structure by product or process in order to make better decisions.

8.3.5.8 Defects per Million Opportunity—Quality Analysis

In Lean Six Sigma, quality is typically measured in defects per million opportunities (DPMO). DPMO provides a standardized apple-to-apples comparison that allows organizations to compare their defect rates by determining the number of defects that would be produced if the company were to produce 1 million products or perform 1 million services. A defect is defined as any nonconformance to customer expectations and includes products that do not function, are damaged in production, are not machined to requirements, or do not match what the customer ordered. The calculation for DPMO is:

$$\frac{(\textit{Total Number of Defects}) * (1{,}000{,}000)}{\textit{Total Number of Opportunities}}$$

For example, if a machine shop produces 55,000 hydraulic cylinders per month and of that total 25 are defective, the DPMO of this machine shop is 454.5. DPMO is typically considered a stretch goal that aims to reduce the number of defects produced to zero.

8.3.5.9 Failure Mode Effects Analysis (FMEA)

Determining the causes of quality failures is a critical early step in improving quality for a product or process. This step helps to identify and quantify quality issues at the source. Failure modes and effects analysis (FMEA) is a step-by-step approach for identifying all possible failures in a design, a manufacturing or assembly process, or a product or service; *failure modes* means the ways, or modes, in which something might fail and *effects analysis* refers to studying the consequences of those failures [9]. Failures are prioritized according to how serious their consequences are, how frequently they occur, and how easily they can be detected; the purpose of the FMEA is to take actions to eliminate or reduce failures, starting with the highest-priority ones [9]. Following is a list of steps required to conduct a FMEA:

1. Form the FMEA team (usually the entire Lean Six Sigma team or subset).
2. Clarify the product or process to be analyzed.
3. Define the purposes (known as functions) of the process or product from the end customer's perspective.
4. Identify all the possible sources of failure modes for each "function"; for example, if a machining process is being analyzed, the team may examine finish issues or dimension issues.
5. List the negative consequences of each failure mode for the end customer.
6. Identify the severity (*S*) of the negative impact for each failure mode using a standardized rating system (typically 1 to 10).
7. Identify the root causes for each failure mode and the probability or occurrence (*O*) of failure mode due to each root cause on a standardized rating system (typically 1 to 10).
8. Identify the current process controls to monitor and prevent each root cause and the ability to detect (*D*) each root cause; rate each on a standardized scale of 1 to 10 to rate the effectiveness of each control.

9. Calculate the risk priority number (RPN) for each failure mode using the following equation: $S \times O \times D$ and the criticality by multiplying severity by occurrence: $S \times O$. These numbers provide guidance for ranking potential failures in the order they should be addressed [9].
10. Target and address the failure modes related to critical characteristics (typically these are failure modes with S ratings for 8, 9, or 10).
11. Identify recommendations and actions to eliminate the failure modes.
12. Record all decisions and actions for future reference to monitor, improve, and control the process and to apply to other processes.

8.3.5.10 Cost of Quality Analysis

After the modes and occurrences of quality failures have been identified, a cost of poor quality analysis can be completed to estimate the financial impact of these quality failures. As mentioned in Section 4.2 earlier the four costs of poor quality are:

1. Prevention
2. Appraisal
3. Internal failure
4. External failure

The Lean Six Sigma team should then group costs related to each category. The goals are to minimize appraisal, internal failure, and external failure costs of the product or service and maximize prevention costs. The team will need to map the process, occurrences, and costs for internal and external failures first. The question, "What happens when an unacceptable project reaches the customer and what are the associated costs?" should be answered. For example, if a machine shop ships an order of 50 components to an automobile supplier with the dimensions outside the acceptable tolerance, the machine shop will have increased costs related to transporting the unacceptable order back to their facility, the cost for employees to process the incorrect order, cost to rework or scrap the unacceptable order, correct the order, and the costs to transport the correct order back to the customer. Once these costs have been calculated, they serve as a strong incentive to the Lean Six Sigma team and organization to minimize them. Similar process maps, occurrence rates, and costs should be calculated for internal failures (orders that are not acceptable, but are identified and not shipped to the end customer), appraisal costs (the costs for the labor and equipment to inspect work in process and end products), and prevention costs. As improvements are made to the process or product, the four costs of quality serve as a benchmark related to cost improvements associated with quality improvements.

8.3.5.11 Process Capability

Process capability measures the nonconformance rate of a process by expressing the performance in the form of a single number and involves calculating the ratio of the specification limits (customer requirements) to process spread (variation in the process). Below is the calculation for the process capability ratio ($\hat{C}_p$):

$$\hat{C}_p = \frac{USL - LSL}{6\sigma}$$

where *USL* and *LSL* are the upper and lower specification limits (customer requirements), respectively, and σ is the standard deviation of the process. Because it is a ratio of the customer requirements over the process variation, a process capability greater than 1.0 is good and indicates that the process is capable of meeting customer requirements. Conversely, a process capability ratio less than 1.0 is bad and indicates that process is not highly capable of meeting customer requirements and the process will generate defects. For example, if an automobile part supplier produces a part that must be 10-cm long with an acceptable range of 9.95 cm and 10.05 cm and the standard deviation of the process is 0.04 cm, the $\hat{C}_p$ is:

$$\hat{C}_p = \frac{10.05 - 9.95}{6(0.04)} = 0.42$$

In this situation, the process is not capable of meeting customer requirements as the variation in the process is greater than the range that the customer will accept ($\hat{C}_p$ is less than 1.0). If the automobile part supplier is able to apply Lean Six Sigma and reduce the process variation to σ = 0.1 cm, the new process capability will be:

$$\hat{C}_p = \frac{10.05 - 9.95}{6(0.01)} = 1.67$$

With an improved process capability greater than 1.0, the process is now capable of meeting customer requirements and will generate few to no defects. In summary, the process capability ratio is an excellent tool to measure the ability of a process to meet customer requirements and provides insights into the amount of defects that will be produced by the process by comparing customer requirements (*USL* and *LSL*) to the process variation (σ).

On a side note, $\hat{C}_p$ assumes that the process is centered on the mean between the *USL* and *LSL*. It is possible that that process is not centered on the mean or shifted from the mean value. If this is the case, the adjusted process capability ratio ($\hat{C}_{pk}$) is:

$$\hat{C}_{pk} = \min\left[\frac{(USL-\mu)}{3\sigma}, \frac{(\mu-LSL)}{3\sigma}\right]$$

where μ is the process mean. If a process is centered on the mean, $\hat{C}_{pk}$ will equal $\hat{C}_p$.

8.3.5.12 Sigma Level Calculation and DPMO Analysis

Determining Sigma levels of processes (One Sigma, Six Sigma, etc.) allows process performance to be compared throughout an entire organization, because it is independent of the process; it is a determination of opportunities and defect rates [10]. The Sigma level relates directly to the defect per million opportunities as discussed in earlier in Section 8.3.5.8. Specifically, the Sigma level based on DPMO is:

One Sigma = 690,000 defects per million units produced.
Two Sigma = 308,000 defects per million units produced.
Three Sigma = 66,800 defects per million units produced.
Four Sigma = 6,210 defects per million units produced.
Five Sigma = 230 defects per million units produced.
Six Sigma = 3.4 defects per million units produced.

The calculation of the Sigma level is based on the process mean and standard deviation using the normal distribution, but the ranges above can be used to estimate a process Sigma level.

8.3.5.13 Pareto Analysis for Improvement Opportunities

After the cost and quality baselines have been established, the final phase of the analysis step is to use a Pareto analysis to identify the most significant and meaningful improvement opportunities that will have the largest impact on the organization. After the baseline data has been calculated, the assessment team can begin to investigate individual quality failures and cost overruns that should be targeted for improvement. A useful method to accomplish this task is to conduct a Pareto analysis. A Pareto analysis is a statistical technique in decision making that is used for the selection of a limited number of tasks that produce the significant overall effect. Pareto analysis is a formal technique useful where many possible courses of action are competing for your attention. In essence, the problem-solver estimates the benefit delivered by each action, then selects a number of the most effective actions that deliver a total benefit reasonably close to the maximal possible one. This analysis uses the Pareto principle—a large majority of effects, in this case quality failures are produced by a few key causes. The Pareto principle is also known as the 80/20 rule that 80% of the effects are caused by 20% of the causes. The idea is to identify the 20% significant quality issues that cause 80% of the total quality failures then target the 20% significant causes for improvement.

8.3.6 Step 6: Identify Major Cost and Quality Improvement Opportunities

After the baseline data have been calculated, the assessment team can begin to investigate individual quality and cost issues that should be targeted for improvement. A useful method to accomplish this task is to review a Pareto analysis. As mentioned previously, a Pareto analysis is a statistical technique in decision making that is used for the selection of a limited number of tasks that produce the significant overall effect. Typically, a spreadsheet and graph are used to conduct the Pareto analysis.

8.3.7 Step 7: Determine, Evaluate, and Select Improvement Alternatives

Once the major opportunities have been identified, the Lean Six Sigma team can begin to develop alternatives to improve quality, reduce costs, and move closer to the established goal of the project based on the team charter. In this phase of the project, the team identifies alternatives to improve quality and reduced cost and then evaluates the economic and operational feasibility while rating each alternative on its ability to achieve the project goals. This section covers the process to generate, screen, and select improvement alternatives. In addition, a comprehensive list of common quality and cost improvement alternatives is provided.

8.3.7.1 Generating Alternatives

Several sources exist to aid in generating alternatives such as existing records review, the Lean Six Sigma results, and the analysis phase. Various methods and tools are available to develop the initial list of alternatives. The environment in which these alternatives should be created is one that encourages creativity and free thinking by the team. Following is a suggested list of methods to identify and create these alternatives:

- Discussions with professional groups and associations
- Discussions with plant engineers and operators
- Internet and literature reviews
- Information available from federal, state, or local governments

- Discussions with equipment manufacturers or vendors
- Discussions with environmental or business consultants
- Brainstorming
- Benchmarking

Trade associations generally provide assistance and information about regulations and the various tools and techniques to address these issues. The information is usually industry specific and at times is free of charge. The National Association of Manufacturers (NAM), the National Association of Purchasing Management (NAPM), and the American Society for Quality (ASQ) are some very good examples of trade associations.

Discussions with plant employees are often a very low cost and very reliable method to develop alternatives to improve quality and reduce costs. Employees are very familiar with a facility's processes and operations. In addition to generating feasible options, these discussions also aid in fostering employee buy-in for the program and increase support. Establishing focus groups, town hall meetings, service talks, facility postings, or one-on-one interviews are methods to gather feedback from employees.

Literature and Internet reviews are another low-cost method to generate alternatives. These include Internet searches, journal reviews, technical magazine reviews, and government reports. These sources describe business processes that other organizations have successfully improved including the methods, tools, and equipment used. Oftentimes, a company does not need to reinvent the wheel if another company successfully improved a process and reported the results. Many articles also offer contact information of the authors to gather information. Some examples of popular environmental magazines and journals are the *International Journal of Lean Six Sigma*, *Competitive Advantage*, and the *International Journal of Six Sigma*.

The government is also a great source of free information and guidebooks. The National Network for Manufacturing Innovation (NNMI) and local governments have developed programs that include technical assistance and information on industry-specific process improvement tools and techniques. These are available at no cost.

Equipment manufacturers and vendors are another source of information to generate alternatives. The downside to this is a bias toward the vendor's own products. Nonetheless, talking to vendors and manufacturers is a good method to identify equipment-related options and they can provide installation assistance.

Consultants are yet another source to generate alternatives. A consultant with Lean Six Sigma experience and industry-specific knowledge can be a very valuable asset to an organization's improvement efforts. The downside is the added cost to contract with a consultant, because the information will rarely be given free of charge.

Team brainstorming sessions are another alternative. Brainstorming is a group creativity technique designed to generate a large number of ideas for the solution to a problem. Throughout the early stages of the project, the team will have to answer several "what," "why," and "how" questions. Brainstorming can accomplish this goal. The general brainstorming rules are listed below:

- Collect as many ideas as possible from all participants with no criticisms or judgments made while ideas are being generated.
- All ideas are welcome no matter how silly or far out they seem. Be creative. The more ideas the better because at this point you don't know what might work.
- Absolutely no discussion takes place during the brainstorming activity. Talking about the ideas will take place after brainstorming is complete.
- Do not criticize or judge. Don't even groan, frown, or laugh. All ideas are equally valid at this point.
- Do build on others' ideas.
- Do write all ideas on a flipchart or board so the whole group can easily see them.
- Set a time limit for the brainstorming, such as 30 minutes or one hour.

Benchmarking may also be used to generate ideas. Benchmarking is defined as the concept of discovering the best performance being achieved, whether within the team's company, by a competitor, or by an entirely different industry. Benchmarking is an improvement tool whereby a company measures its performance or process against other companies' best practices, determines how those companies achieved their performance levels, and uses the information to improve the team's company performance. Data for benchmarking can be achieved from company visits, annual reports, or Internet searches. Oftentimes, team members have colleagues in other companies and industries that can provide these data. Data can also be collected from regional committees. For example, a multiorganization Round Table Committee was formed by a local nonprofit group to meet monthly to discuss challenges and successes. At the monthly meeting, lunch is provided and new members are welcome at no cost. Also at the meeting, each member provides a status update of his or her company's efforts, process changes, and issues where the company could use assistance.

Each waste improvement alternative should be documented to facilitate the screening and review process. A simple spreadsheet that records the name of the alternative, a brief description, the author with contact information, and sources (if any) is a very useful tool. Figure 8.20 is a sample spreadsheet that may be used to facilitate this process.

	Alternative	Description	Author	Contact Information (phone/e-mail)	Sources or Additional Information
1					
2					
3					
4					
5					
6					
7					
8					
9					
10					
11					
12					
13					
14					
15					
16					
17					
18					
19					
20					

FIGURE 8.20
Alternative generation worksheet.

8.3.7.2 Screening Alternatives

The process of creating cost and quality alternatives can generate numerous options. It would be very time consuming for the team to conduct detailed financial and operational feasibility evaluations on each option. A fast screening process can help to identify quickly the options worthy of full evaluations and possible inclusion in the Lean Six Sigma program. In addition, ineffective options can be quickly weeded out, saving the team time and money in the evaluation process. An effective screening process should be based on the original goals of the project and at a minimum should examine:

- Expected quality improvement
- Expected startup costs
- Impact on operational costs ($ per year)
- Impact on purchasing costs ($ per year)
- Impact on employee morale
- Ease of implementation

The team should keep in mind that the goal of the screening process is to identify quickly options worthy of further analysis. A weighed scoring system can be developed to rank each alternative consistently in an objective manner. A quality deployment function (QDF), such as a House of Quality is an excellent tool to accomplish this evaluation. A House of Quality is a graphic tool for defining the relationship between an organization's desires and the organization's capabilities. It utilizes a planning matrix to relate the organizational wants (e.g., quality and cost improvements) to how the Lean Six Sigma program will or can meet those wants (e.g., process changes). It looks like a house with a correlation matrix as its roof and organizational wants versus waste minimization options as the house. The House of Quality can also increase cross-functional integration within organizations using it, especially between marketing, engineering, and manufacturing.

The basic structure is a table with the rows on the left labeled as What and the columns as How. Rankings based on the Whys and the correlations can be used to calculate priorities for the Hows. House of Quality analysis can also be cascaded, with Hows from one level becoming the Whats of a lower level; as this progresses the decisions get closer to the engineering/manufacturing details.

Before proceeding with the screening process, the team should decide on the evaluation criteria (the Whats) and weighting system. A scale of 1–10 for weighting each criterion is recommended. These weightings should be determined by the team, project manager, facility manager, or a combination of all three. The evaluation criteria should be directly related to the overall goals of the project, such as:

- Productivity improvements
- Quality improvement
- Reduction in purchasing costs
- Revenue generation potential
- Low startup costs
- Reduction in waste amounts
- Ease of implementation
- Impact on employee morale
- Impact on organization image
- Impact on safety
- Other factors as determined by the team

Once these criteria have been created, the team should rank them on a scale of 1–10 based on importance. For example, quality and lead time improvements could receive an importance rating of 10 (meaning it is highly important) versus low startup costs receiving an importance rating of 2 (meaning that startup costs are of low importance and not a major factor in the decision process). These criteria should then be placed into the column headers of the spreadsheet with importance weightings in parentheses. Figure 8.21 provides a template for screening alternatives.

Option	Productivity improvements	Quality improvement	Reduction in purchasing costs	Revenue generation potential	Low startup costs	Reduction in waste amounts	Ease of implementation	Impact on employee morale	Impact on organization image	Impact on safety	**Score**

FIGURE 8.21
Alternative screening worksheet.

Once the criterion and importance ratings have been established, the team should list each alternative in the rows under the Option column. In the row for each alternative, the team should place a rating score corresponding to the level of which the alternative meets the criterion with 0 being no impact and 10 being great impact. For example, if the team is considering the purchase of a new, more accurate machine, the quality improvements could be significant, so the team could rate it an 8, but in the startup cost criterion, the team could rate it lower, such as a 1, due to the high implementation cost. Once each alternative is rated, the ratings should be multiplied by the importance factor and each row should be summed. This score will allow the team to screen each alternative objectively. Once all of the alternatives are listed and scored, the team can screen them based on their total score. Alternatives with higher total scores pass the screening process and become eligible for further evaluation. To determine the cutoff point, depending on time and money resources, the team could set the threshold at a specific point value, accept the top 20%, or accept the top 10 for further analysis. When first starting a Lean Six Sigma program it is recommended that the team select the top third of all alternatives for further screening to compensate for estimation error. Figure 8.22 provides an example of a completed screening worksheet.

As displayed in the example in Figure 8.22, the process change received a higher score (399) versus the purchase of a new machine (371). This score

Option	*Productivity improvements (10)*	*Quality improvement (10)*	*Reduction in purchasing costs (7)*	*Revenue generation potential (8)*	*Low startup costs (7)*	*Reduction in waste amounts (4)*	*Ease of implementation (6)*	*Impact on employee morale (6)*	*Impact on organization image (5)*	*Impact on safety (10)*	***Score***
New machine	10	9	2	2	2	6	3	5	1	6	371
Process change	3	5	4	2	10	6	8	8	1	8	399

FIGURE 8.22
Completed alternative screening worksheet.

difference indicates that the process change better meets the overall needs of the organization versus the purchase of a new machine and that the process change should be given preference over the machine purchase.

8.3.7.3 Analyzing and Selecting Alternatives

After trimming down the list of alternatives via the screening process, the remaining alternatives should be further analyzed to determine the best fit for the organization to implement Lean Six Sigma and hence include in the program. The analysis process focuses on identifying the benefits, costs, and drawbacks of each alternative. To accomplish this, each alternative is evaluated based on:

- Impact on the program goal
- Technical feasibility
- Operational feasibility
- Economic feasibility
- Sustainability
- Organizational culture feasibility

The key outcome of this phase is to fully document, analyze, and arrive at a final acceptance decision for each alternative. To accomplish this, process flowcharts are analyzed, a complete feasibility analysis is completed (including technical, operational, and organizational), a cost justification study is completed, feedback is collected and analyzed, and finally a decision made regarding each alternative. This feasibility analysis provides a complete discussion and documentation of each alternative that will be used in the implementation phase if the alternative is accepted. During this process the team must keep a clear understanding of the overriding goals of the Lean Six Sigma project and the organization, for example, the relative importance of reducing costs versus improving quality. Some alternatives may require extensive analysis, including gathering additional data from vendors or analyzing market trends for markets. The first consideration when evaluating alternatives is the impact on the goals of the project established in the first phase of the project. Alternatives should be separated into different categories to aid with this process.

The evaluation process itself consists of seven steps to rate each alternative. The process is completed sequentially and after the first step (process description) and after each following step, the alternative is accepted to move to the next phase or rejected and the analysis is terminated. If the alternative does not meet thresholds or feasibility tests it is eliminated from further review to save the team time and resources. The alternative should still be kept on file in the event technology or organizational changes render the option feasible. Below is a summary of the seven-step evaluation process:

1. Fully describe each alternative in terms of the equipment, raw material, process, or purchasing additions or modifications.
2. Calculate the annualized cost, quality, and customer satisfaction impacts.
3. Compile and analyze the process flowcharts related to each process change.
4. Conduct feasibility analyses (technical, operational, and organizational).
5. Conduct a cost justification for each alternative (payback, internal rate of return, net present value).
6. Gather feedback from all stakeholders.
7. Get approval and signoff from the Lean Six Sigma team and executives.

Technical and operational feasibility is concerned with whether the proper resources exist or are reasonably attainable to implement a specific alternative. This includes the square footage of the building, existing and available utilities, existing processing and material handling equipment, quality requirements, and skill level of employees. During this process, product specifications and facility constraints should be taken into account. Typically technical evaluation criteria include:

- Available space in the facility
- Safety
- Compatibility with current work processes and material handling
- Impact on product quality
- Required technologies and utilities (power, compressed air, data links)
- Knowledge and skills required to operate and maintain the alternative
- Additional labor requirements
- Impact on product marketing
- Implementation time

When evaluating technical feasibility, the facility engineers and consultants (if utilized) should be contacted for input. In addition it is also wise to discuss the technical aspects with workers directly affected by the change such as production and maintenance. If an alternative calls for a change in raw materials, the effect on the quality of the final project must be evaluated. If an alternative does not meet the technical requirements of the organization, it should be removed from consideration. From a technical standpoint, the three areas that require additional evaluation are:

- Equipment modifications or purchases
- Process changes
- Material changes

If an alternative involves an equipment modification or purchase, an analysis of the equipment should be conducted. The team should investigate whether the equipment is available commercially and the contact from the manufacturer. Performance of the machine should also be addressed, including cost, utility requirements, capacity, throughput, cycle time, required preventive maintenance, space requirements, and possible locations in the facility where the equipment could be installed. In addition, if production would be affected during installation, this should be evaluated as well. The vendor or manufacturer could provide more information regarding potential shutdowns. Required modifications to workflow or production procedures should be analyzed as well as required training and safety concerns related to the equipment purchase or modification. From an operational standpoint, attention should be given to how the alternative will improve or reduce productivity and labor force reductions or increases.

If an improvement alternative involves a process change or a material change, the affected areas should be identified and feedback gathered from the area managers, employees, maintenance, and engineers (if applicable). With the process changes, training requirements should also be discussed. Also, the impacts on production, material handling/storage, and quality should be addressed. A material testing program is highly recommended for new items that the engineering team may not be familiar with to analyze quality and throughput impacts. A design of experiment (DOE) that tests the changes versus the current material is an excellent method to gauge impacts. A DOE involves collecting data to better understand the causes of variation in an experiment. These causes can be controllable, such as the temperature of an injection press; or uncontrollable, such as naturally occurring variation in raw material. Often the experimenter is interested in the effect of some process or intervention, such as using a new raw material, on some outcome such as quality.

From an economic standpoint, traditional financial evaluation is the most effective method to analyze alternatives. These measures include the pay back period (discounted payback period), internal rate of return, and net present value for each alternative. If the organization has a standard financial evaluation process, this should be completed for each alternative. The accounting or finance department would have this information. To perform these financial analyses, revenue and cost data must be gathered and should be based on the expectations for the alternatives. This is more complicated than it sounds, especially if a project will have an impact on the number of required labor hours, utility costs, and productivity, not to mention initial investments. A comprehensive estimation of the cost impacts (revenues and costs) per year over the life of the alternative is required to begin the analysis. The first step of the economic evaluation process is to determine these costs. These costs include capital costs (or initial investment), operating costs/savings, operating revenue, and salvage values for each waste minimization alternative.

Capital costs are the costs incurred when purchasing assets that are used in production and service. Normally they are nonrecurring and used to purchase large equipment such as a baler or plastic grinder. Capital costs include more than just the actual cost of the equipment; they also include the costs to prepare the site for production. Following is a brief list of typical capital costs, also known as the initial investment:

- Site development and preparation (including demolition and clearing if needed)
- Equipment purchases including spare parts, taxes, freight, and insurance
- Material costs (piping, electrical, telecommunications, structural)
- Building modification costs (utility lines, construction costs)
- Permitting costs, building inspection costs
- Contractor's fees
- Startup costs (vendor, contractor, in-house)
- Training costs

After the initial investment has been calculated, the recurring costs, savings, and revenues from the improvement alternative must be determined. The concept is to improve quality and to reduce operational costs based on the implementation of the alternative under analysis. For example, if a company considers the installation of a new drill press, the annual operating costs of the drill press (such as labor and utilities), the annual cost savings from reduced labor costs, and the revenue from increased sales should be considered. Some common recurring costs include:

- *Changes in utility costs:* Utility costs may increase or decrease depending on the installation, modification, or removal of equipment.
- *Changes in operating and maintenance labor/benefits:* An alternative may increase or decrease labor requirements and the associated benefits. They may be reflected in changes in overtime hours or in the number of employees.
- *Changes in operating and maintenance supplies:* An alternative may result in increased or decreased operating and maintenance supply usage.
- *Changes in overhead costs:* Large projects may increase or decrease these values.
- *Changes in revenues for increased (or decreased) production:* An alternative may result in an increase in the productivity of a unit. This will result in changes in revenue.

- *Input material cost savings:* Options that reduce scrap, reduce waste, or increase internal recycling tend to decrease the demand for input materials.
- *Increase revenue from by-products:* An alternative may generate a by-product that can be sold to a recycler or sold to another company as material. This will increase a company's revenue.
- *Reduced solid waste disposal costs:* Waste generation is reduced or is diverted to recycling streams resulting in less waste being sent to the landfill for disposal and lower hauler charges. These include disposal fees, transportation costs, and predisposal treatment costs.

It is suggested that savings in these costs be taken into consideration first, because they have a greater impact on the project economics and involve less effort to estimate reliably. The remaining elements usually have a smaller impact and should be included on an as-needed basis or to fine-tune the analysis.

A project's profitability is measured by estimating the net cash flows each operating year over the life of the project. A net cash flow is calculated by subtracting the cash outlays from the cash incomes starting in year zero (the year the project is initiated). Figure 8.23 is an example of a cash flow diagram.

If a project does not have an initial investment, the project's profitability can be judged by whether an operating cost savings occurred. If such a project reduces overall operating costs, it should be implemented. For example, if a manufacturing company is able to modify a standard operating procedure and reduce cycle times by adding a new worker to the process, there is little to no initial investment with the example, but there will be added labor costs for the new worker versus additional revenue generated by the worker. If the additional revenues outweigh the additional costs, the alternative should be implemented.

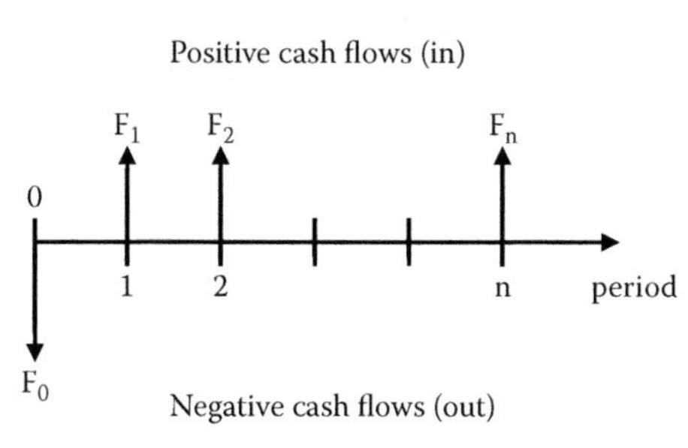

FIGURE 8.23
Cash flow diagram example.

For projects with significant initial investments or capital costs, a more detailed profitability analysis is needed. The three standard measures of profitability are:

- Payback period
- Internal rate of return (IRR)
- Net present value (NPV)

The payback period for a project is the amount of time it requires to recover the initial cash outlay for the project. The formula for calculating the payback period on a pretax basis in years is:

$$\text{Payback Period} = \frac{\text{Capital Investment}}{\text{Annual operating cost savings}}$$

For example, suppose a manufacturer installs a new, faster punch press for a total cost of \$65,000. If the punch press is expected to save the company \$20,000 per year, then the payback period is 3.25 years. Payback period is typically measured in years. However, some alternatives may have payback periods in terms of months. Many organizations use the payback period as a screening method before conducting a full financial analysis. If the alternative does not meet a predetermined threshold, the alternative is rejected. Payback periods in the range of three to four years are usually considered acceptable for low-risk investments. Again, this method is recommended for quick assessments of profitability. If large capital expenditures are involved, it should be followed by a more strenuous financial analysis such as the IRR and NPV.

The internal rate of return and net present value are both discounted cash flow techniques for determining profitability and determining if a Lean Six Sigma alternative will improve the financial position of the company. Many organizations use these methods for ranking capital projects that are competing for funds, such as the case with Lean Six Sigma alternatives. Capital funding for a project can depend on the ability of the project to generate positive cash flows beyond the payback period to realize an acceptable return on investment. Both the IRR and NPV recognize the time value of money by discounting the projected future net cash flows to the present. For investments with a low level of risk, an after-tax IRR of 12 to 15% is typically acceptable. The formula for NPV is:

$$NPV = \sum_{t=0}^{N} \frac{C_t}{(1+r)^t}$$

Notice that each cash inflow/outflow is discounted back to its present value and then is summed; where

t = the time of the cash flow.
N = the total time of the project.
r = the discount rate (the rate of return that could be earned on an investment in the financial markets with similar risk).
C_t = the net cash flow (the amount of cash) at time t (for educational purposes, C_0 is commonly placed to the left of the sum to emphasize its role as the initial investment).

The internal rate of return is a capital budgeting metric used by firms to decide whether they should make investments. It is an indicator of the efficiency of an investment, as opposed to net present value, which indicates value or magnitude. The IRR is the annualized effective compounded return rate that can be earned on the invested capital, that is, the yield on the investment.

A project is a good investment proposition if its IRR is greater than the rate of return that could be earned by alternate investments (investing in other projects, buying bonds, even putting the money in a bank account). Thus, the IRR should be compared to any alternate costs of capital including an appropriate risk premium. Mathematically the IRR is defined as any discount rate that results in a net present value of zero of a series of cash flows. In general, if the IRR is greater than the project's cost of capital, or hurdle rate, the project will add value for the company. The equation for IRR is:

$$NPV = \sum_{t=0}^{N} \frac{C_t}{(1+r)^t} = 0$$

Most spreadsheet programs typically have the ability to calculate IRR and NPV automatically from a series of cash flows. Following is an example applying these financial evaluation concepts. Returning to the new punch press example discussed previously, recall an initial cost of \$65,000 and \$20,000 in annual savings and assume a punch press life span of 10 years and an organization minimum attractive rate of return (MARR) of 15%. The MARR is the minimum return on a project that a manager is willing to accept before starting a project, given its risk and the opportunity cost of forgoing other projects. Table 8.3 displays the cash flows, IRR, and the NPV result.

As shown in the last two rows of Table 8.3, the IRR is 28.2% and the NPV is nearly \$31,000 at a MARR of 15%. The fact that the IRR is greater than the 15% MARR and the fact that the NPV is positive indicates that the project is a good financial decision.

Lean Six Sigma alternatives should also be evaluated based on sustainability and the cultural fit within the organization. *Sustainability* is defined as an organization's investment in a system of living, projected to be viable on an ongoing basis that provides quality of life for all individuals. Sustainability in its simplest form describes a characteristic of a process or state that can

TABLE 8.3

Net Present Value Analysis

Year	Cash Flow
0	$(65,000)
1	$20,000
2	$20,000
3	$20,000
4	$20,000
5	$20,000
6	$20,000
7	$20,000
8	$20,000
9	$20,000
10	$20,000
IRR	28.2%
NPV	$30,761

be maintained at a certain level indefinitely. The term, in its environmental usage, refers to the potential longevity of vital human ecological support systems, such as the planet's climatic system, systems of agriculture, industry, forestry, fisheries, and the systems on which they depend. In other words, the Lean Six Sigma alternatives should be evaluated based on how well they meet this definition, such that the alternative can be sustained without large amounts of effort or additional resources and continue to protect the environment. Often, this will be related to the culture of the organization. Criteria commonly used to evaluate the sustainability of an alternative include:

- Dealing transparently and systemically with risk, uncertainty, and irreversibility
- Ensuring appropriate valuation, appreciation, and restoration of nature
- Integration of environmental, social, human, and economic goals in policies and activities
- Equal opportunity and community participation/sustainable community
- Conservation of biodiversity and ecological integrity
- Ensuring intergenerational equity
- Recognizing the global integration of localities
- A commitment to best practice
- No net loss of human capital or natural capital
- The principle of continuous improvement
- The need for good governance

When an alternative involves working with a third-party vendor there are several key questions to ask potential candidates to determine the best fit for the organization. Those questions include:

- What contract terms does the vendor require?
- Who provides the transportation if applicable?
- What is the schedule of transportation if applicable?
- Are there minimum quantity requirements?
- Who will provide containers for work in process or finished goods?
- Can "escape clauses" be included in the contract?
- Can the vendor provide references?

In a similar vein, when working with equipment vendors, there several key questions to ask:

- What is the total cost of the equipment including freight and installation?
- What are the building requirements and specifications for the equipment (compressed air, electricity, space, minimum door widths)?
- Is a service contract included in the purchase price or is there an additional charge?
- Do you offer training to the employees, engineers, and maintenance employees who will be working with the equipment? If so, is there a charge?
- What is the process if the equipment malfunctions and the company needs support? Is there a representative available 24 hours per day? What is the charge for these visits?
- Do you offer an acceptance test process to ensure that the equipment operates within the promised specifications (capacity and cycle time)?
- What is the required installation time and must production be shut down?

The following worksheets may be used to evaluate each Lean Six Sigma alternative. The first worksheet is a cover page that provides a general description of the alternative. Based on the responses to the questions on the cover sheet one of four additional worksheets is attached (see Figures 8.24 to 8.32):

- Worksheet A: Equipment Purchases or Modifications
- Worksheet B: Raw Material Changes
- Worksheet C: Process Changes
- Worksheet D: Purchasing Changes

Lean Six Sigma Alternative Feasibility Analysis Worksheet
Cover Sheet
The purpose of this work sheet is to evaluate the feasibility of alternatives that have passed the screening process

Company Name ____________________
Location: ____________________
Date: ____________________

Alternative Description:

Alternative Tracking Number: date and two digit number
Work Unit of Departments Affected:

Source or Contact Person:
Contact Information:

Does the Alternative Involve (check the ONE that applies):

☐ Equipment Purchases or Modifications > **Please Attach Worksheet A**
☐ Raw Material Changes > **Please Attach Worksheet B**
☐ Process Changes > **Please Attach Worksheet C**
☐ Purchasing Changes > **Please Attach Worksheet D**

FIGURE 8.24
Feasibility analysis coversheet.

The worksheets are designed to work together to describe and evaluate each Lean Six Sigma alternative completely. If the alternative is approved, the form also provides signoffs for the leadership team to indicate review and acceptance. Section III of this book provides 11 case studies that apply the alternative evaluation process and worksheets with accompanying data.

8.3.8 Step 8: Develop Lean Six Sigma Deployment and Execution Plan

The goal of this step of the Lean Six Sigma process is to translate the list of accepted alternatives from Step 7 into an achievable implementation plan and to document the selected alternatives. The remainder of this section provides an overview of the deployment plan, a discussion regarding obtaining funding, and details regarding each of the sections of a deployment plan. Finally, a template for the deployment plan is provided.

Lean Six Sigma Alternative Feasibility Analysis Worksheet A
Equipment Purchases or Modifications

The purpose of this work sheet is to evaluate the feasibility of alternatives that have passed the screening process. All grayed boxes should be completed before completing the Approval Process in Step 6

Company Name ______________________

Location: ______________________

Date: ______________

Alternative Description: ______________________

Alternative Tracking Number: ______

			PASS
Step 1: Estimate quality improvements and attach supporting documents		%	
Estimate productivity improvements		parts/year	
Estimate cost reductions		$/year	
Estimate revenue increase	$	$/year	
Step 2: Attach process flowcharts for the process			
Step 3: Feasibility Analysis			
Technical			
Does the equipment exist (vendor or manufacturer)?	Yes / No		
Description of Machine			
Vendor Name			
Vendor Contact Information			
Estimate Machine and Installation Cost	$		
Implementation Time (days)		days	
Required utilities			
Power	Yes / No		
Compressed air	Yes / No		
Data link	Yes / No		
Other			
Compatibility with current work processes and material handling			
Training Concerns			
Skill level required to operate equipmetn			
Skill level required to maintain equipmetn			
Space Requirements			
Space required for machine and staging		ft^2	
Available space in the facility		ft^2	
Proposed locations			
Operational			
Machine cycle time		minutes	
Machine capacity		units	
Labor impacts (Additional work hours)			
Supervisory needs			
Maintenance needs			
Productivity impacts			
Safety concerns and impacts			
Product quality impact			
Addition labor requirements			
Organizational			
Impact on sales			
Impact on marketing			
Impact on employee morale			
Impact on corporate image			
Impact on supply chain			

FIGURE 8.25
Feasibility analysis worksheet A—page 1 of 2.

Lean Six Sigma Alternative Feasibility Analysis Worksheet A (Page 2 of 2)
Equipment Purchases or Modifications

Step 4 Cost Analysis

Machine Costs

Machine cost	$
Site development	$
Material Costs	$
Building modification costs	$
Permitting and inspection costs	$
Contractor fees	$
Start up costs	$
Initial training costs	$

Operating Expenses

Utility cost impacts	$
Input material changes	$
Labor cost impacts	$
Supervision cost impacts	$
Maintenance cost impacts	$
Operating and maintenance supply impacts	$
Changes in Overhead costs	$

Operating Savings and Revenue

Reduced operating costs	$
Revenues from increased productivity	$
Revenues from the sale of end products	$

Life of Machine or product (whichever is shorter)	years

Total Initial Investment	$
Annual Operating Savings	$
Payback period	$
Net Present Value (NPV)	$
Internal Rate of Return (IRR)	
Organization Minimum Attractive Rate of Return (MARR)	

Step 5 Feedback Analysis

Feedback from operators

Feedback from management

Feedback from maintenance

Feedback from finance

Step 6 Approval

Lean Six Sigma Team Leader	name	date
Manager, Maintenance	name	date
Manager, Operations	name	date
Manager, Finance	name	date
CEO	name	date

FIGURE 8.26
Feasibility analysis worksheet A—page 2 of 2.

8.3.8.1 Overview of Deployment Plan

A deployment plan is a comprehensive document that details the what, when, where, and how of each alternative. It serves as an implementation guide to aid the organization in achieving its Lean Six Sigma goals. The deployment plan describes the set of tasks necessary to implement a program so that

Lean Six Sigma Alternative Feasibility Analysis Worksheet B
Raw Material or Modifications

The purpose of this work sheet is to evaluate the feasibility of alternatives that have passed the screening process.
All grayed boxes should be completed before completing the Approval Process in Step 6

Company Name ____________________
Location: ____________________
Date: ____________
Alterative Description: ____________________
Alternative Tracking Number: ___

			PASS
Step 1: Estimate quality improvements and attach supporting documents		%	
Estimate productivity improvements		parts/year	
Estimate cost reductions		$/year	
Estimate revenue increase	$	$/year	
Step 2: Attach process flowcharts that generate waste stream			
Step 3: Feasibility Analysis			
Technical			
Does the material exist and is it available	Yes / No		
Compatibility with current work processes and material handling			
Operational			
Labor Impacts (Additional work hours)			
Supervisory Needs			
Maintenance Needs			
Productivity Impacts			
Safety Concerns and Impacts			
Product Quality Impact			
Addition labor requirements			
Organizational			
Impact on sales			
Impact on marketing			
Impact on employee morale			
Impact on corporate image			
Impact on Supply Chain			

FIGURE 8.27
Feasibility analysis worksheet B—page 1 of 2.

it can be effectively transitioned within the organization. The deployment plan provides a detailed schedule of events, persons responsible, and event dependencies required to ensure successful cutover to the new system.

Deployment of a Lean Six Sigma plan can impose a great deal of change and stress on the employees within an organization. Therefore, ensuring a smooth transition is a key factor in the implantation process. The deployment plan should minimize the impact of the implementation on the organization's staff, production system, and overall business routine. The Lean Six Sigma team leader usually is responsible for creating the deployment plan and it is implemented by the entire organization, so communication and feedback are critical in the development process. The Lean Six Sigma deployment plan consists of 10 sections:

1. Cover page with official approvals
2. Overview
3. Assumptions, Dependencies, Constraints
4. Operational Readiness

Lean Six Sigma Alternative Feasibility Analysis Worksheet B (Page 2 of 2)
Raw Material or Modifications

Step 4: Cost Analysis ☐

New Material Costs

Previous Material cost per load	$
New material cost per load	$
Material cost differential	$
Annual loads purchased	$
Annual material cost differential	$

Operating Expenses

Utility cost impacts	$
Labor cost impacts	$
Supervision cost impacts	$
Maintenance cost impacts	$
Operating and maintenance supply impacts	$
Changes in Overhead costs	$

Operating Savings and Revenue

Reduced operating costs	$
Revenues from increased productivity	$
Revenues from the sale of end products	$

Total Initial Investment	$
Annual Operating Savings	$
Payback period	$
Net Present Value (NPV)	$
Internal Rate of Return (IRR)	
Organization Minimum Attractive Rate of Return (MARR)	

Step 5: Feedback Analysis ☐

Feedback from operators	
Feedback from management	
Feedback from maintenance	
Feedback from finance	

Step 6: Approval ☐

Lean Six Sigma Team Leader	name	date
Manager, Maintenance	name	date
Manager, Operations	name	date
Manager, Finance	name	date
CEO	name	date

FIGURE 8.28
Feasibility analysis worksheet B—page 2 of 2.

5. Timeline for Implementation
6. Training and Documentation
7. Notification of Deployment
8. Operations and Maintenance Plan
9. Contingency Plan
10. Appendices

Lean Six Sigma Alternative Feasibility Analysis Worksheet C
Process Changes

The purpose of this work sheet is to evaluate the feasibility of alternatives that have passed the screening process.
All grayed boxes should be completed before completing the Approval Process in Step 6

Company Name ____________
Location: ____________
Date: ____________
Alterative Description: ____________
Alternative Tracking Number: ____

PASS

Step 1: Estimate quality improvements and attach supporting documents %
Estimate productivity improvements parts/year
Estimate cost reductions $/year
Estimate revenue increase $ $/year

Step 2: Attach process flowcharts for the process

Step 3: Feasibility Analysis

Technical
- Compatibility with current work processes and material handling

Operational
- Labor Impacts (Additional work hours)
- Supervisory Needs
- Maintenance Needs
- Productivity Impacts
- Safety Concerns and Impacts
- Product Quality Impact
- Addition labor requirements

Organizational
- Impact on sales
- Impact on marketing
- Impact on employee morale
- Impact on corporate image
- Impact on Supply Chain

FIGURE 8.29
Feasibility analysis worksheet C—page 1 of 2.

8.3.8.2 Contents of Deployment Plan

This section provides an in-depth discussion of the contents of the deployment plan. One of the most important components of the plan is the cover sheet, which contains the official signatures of the executive management team indicating full approval of the deployment plan. Following are the specific contents:

- *Cover sheet:* In addition to the official approvals, the cover page should include the company name, the title of the program (such as Lean Six Sigma Deployment Plan), the date the report was written, and the author or department that created the report. If the organization uses specialized tracking or identification numbers for projects, these should be included too.
- *Overview:* This section provides the purpose, business context, and project summary in an executive summary format. It identifies the purpose of the deployment plan and its intended audience, describes the business processes that will be modified as a result of the deployment, and provides a summary of the plan. It also includes

Lean Six Sigma Alternative Feasibility Analysis Worksheet C (Page 2 of 2)
Process Changes

Step 4: Cost Analysis

New Material Costs

Previous Material cost per load	$
New material cost per load	$
Material cost differential	$
Annual loads purchased	$
Annual material cost differential	$

Operating Expenses

Utility cost impacts	$
Labor cost impacts	$
Supervision cost impacts	$
Maintenance cost impacts	$
Operating and maintenance supply impacts	$
Changes in Overhead costs	$

Operating Savings and Revenue

Reduced operating costs	$
Revenues from increased productivity	$
Revenues from the sale of end products	$

Total Initial Investment	$
Annual Operating Savings	$
Payback period	$
Net Present Value (NPV)	$
Internal Rate of Return (IRR)	
Organization Minimum Attractive Rate of Return (MARR)	

Step 5: Feedback Analysis

Feedback from operators

Feedback from management

Feedback from maintenance

Feedback from finance

Step 6: Approval

Lean Six Sigma Team Leader	name	date
Manager, Maintenance	name	date
Manager, Operations	name	date
Manager, Finance	name	date
CEO	name	date

FIGURE 8.30
Feasibility analysis worksheet C—page 2 of 2.

an overview of activities necessary to get the program launched into the business environment such as installation, configuration, and initial operational activities. It also includes details regarding the location where the assessment was conducted, the dates that it was conducted, the names of the individuals conducting the project, and a facility layout.

- *Assessment Findings and Recommendations:* This section provides a summary of the principal findings, recommendations, and

Lean Six Sigma Alternative Feasibility Analysis Worksheet D
Purchasing Changes

The purpose of this work sheet is to evaluate the feasibility of alternatives that have passed the screening process. All grayed boxes should be completed before completing the Approval Process in Step 6

Company Name ____________________
Location: ____________________
Date: ____________
Alterative Description: ____________________
Alternative Tracking Number: ____

PASS

Step 1: Estimate quality improvements and attach supporting documents %
Estimate productivity improvements parts/year
Estimate cost reductions $/year
Estimate revenue increase $ $/year

Step 2: Attach process flowcharts for the process

Step 3: Feasibility Analysis

Technical

Does the material exist (vendor or manufacturer)? Yes No

Compatibility with current work processes and material handling

Operational

Safety Concerns and Impacts

Product Quality Impact

Addition labor requirements

Organizational

Impact on sales

Impact on marketing

Impact on employee morale

Impact on corporate image

Impact on Supply Chain

FIGURE 8.31
Feasibility analysis worksheet D—page 1 of 2.

observations. It discusses the data collected during the assessment in terms of quality and cost improvements. It also provides a listing of the approved alternatives for implementation.

- *Assumptions, Dependencies, and Constraints:* This section describes the assumptions about the current organizational capabilities and the day-to-day operations of the program. In addition it describes the dependencies that can affect the deployment of the program, such as working within the constraints of third-party vendors, staff expertise, and the factors that limit the ability to deploy the program.
- *Operational Readiness:* This section describes the preparation required for the site on which the program and alternatives will operate. It defines any changes that must occur to the operational site and specifies features and items that should be modified to adapt to the program alternatives. It also describes the method for use in assessing deployment readiness and identifies the configuration audits and reviews to be held after the program is tested and accepted and before the program or equipment is installed in the production environment.
- *Timeline for Implementation:* This section describes the timetable for the implementation of each Lean Six Sigma project or the entire program. It serves as the control document to facilitate communication

Lean Six Sigma Alternative Feasibility Analysis Worksheet D (Page 2 of 2)
Purchasing Changes

Step 4: Cost Analysis

New Material Costs	
Previous Material cost per load	$
New material cost per load	$
Material cost differential	$
Annual loads purchased	$
Annual material cost differential	$
Operating Expenses	
Utility cost impacts	$
Labor cost impacts	$
Supervision cost impacts	$
Maintenance cost impacts	$
Operating and maintenance supply impacts	$
Changes in Overhead costs	$
Operating Savings and Revenue	
Reduced operating costs	$
Revenues from increased productivity	$
Revenues from the sale of end products	$
Total Initial Investment	$
Annual Operating Savings	$
Payback period	$
Net Present Value (NPV)	$
Internal Rate of Return (IRR)	
Organization Minimum Attractive Rate of Return (MARR)	

Step 5: Feedback Analysis

Feedback from operators	
Feedback from management	
Feedback from maintenance	
Feedback from finance	

Step 6: Approval

	name	date
Waste Minimization Team Leader		
Manager, Maintenance		
Manager, Operations		
Manager, Finance		
CEO		

FIGURE 8.32
Feasibility analysis worksheet D—page 2 of 2.

within (departments) and outside the organization (suppliers and contractors). A Gantt chart is an excellent diagram to include in this section as it displays the order or precedence of events and the percent completion to the established timeline. Figure 8.33 provides an example of a deployment timeline.

- *Training and Documentation:* This section describes the plans for preparing and conducting training for the purpose of training all stakeholders regarding program or process changes. It also identifies and

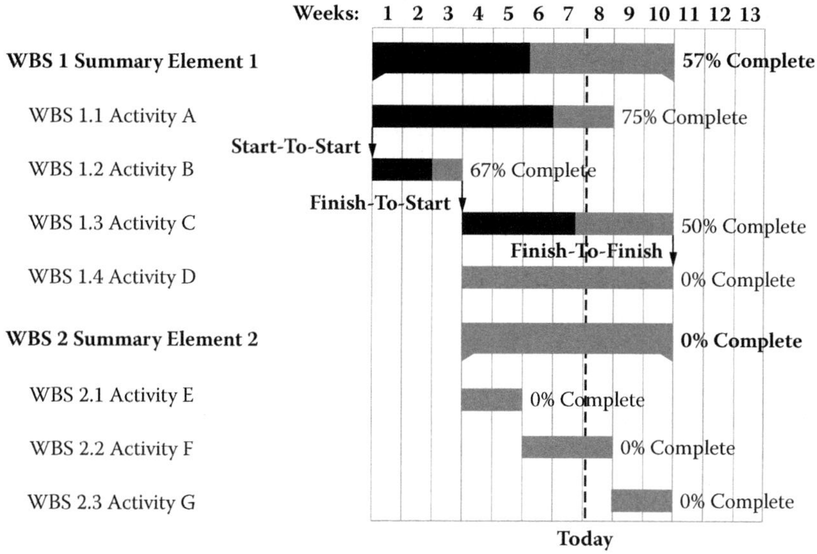

FIGURE 8.33
Sample deployment timeline.

describes each document that will be produced for the purpose of aiding in implementation, support, or use of the new programs. The section should include the activities needed to create each document.

- *Notification of Deployment:* This section describes the method of notifying all stakeholders of the successful release of all Lean Six Sigma programs and identifies stakeholders and groups requiring notification.
- *Operations and Maintenance Plans:* This section describes the maintenance and operations activities for each program or piece of equipment. For example, preventive maintenance schedules should be included for each new piece of equipment.
- *Contingency Plan:* This section describes the contingency plan to be executed if problems occur during deployment activities. A contingency plan is devised for specific situations if or when things do not occur as expected or circumstances change. Contingency plans include specific strategies and actions to deal with specific variances to assumptions resulting in a particular problem, emergency, or state of affairs. They also include a monitoring process and "triggers" for initiating planned actions. They are very useful to help governments, businesses, or individuals recover from serious incidents in the minimum time with minimum cost and disruption.
- *Appendices:* This section contains all relevant appendices related to the project. The alternative evaluation sheets should be included in this section.

8.3.9 Step 9: Execute and Implement Lean Six Sigma Plan

A well-developed deployment plan based on viable options will yield poor results if the plan is not executed properly. There is no such thing as "over-communication" when it comes to rolling out a new project or program. The three key components of a successful implementation and execution are following the deployment plan, communication, and recognizing the need to adjust in certain circumstances.

To facilitate the communication process, at a minimum, weekly progress meetings should be held with all key stakeholders. These meetings should focus on the status of each project versus the timeline and established goals. An agenda and the project timeline should be prepared in advance and serve to lead the discussion. The task leader (as determined in the deployment plan) should take the lead role in discussing the status of each project or program. Any obstacles or delays should be discussed so that the team may determine solutions.

During the deployment process it is critical not to overwhelm employees with process changes. Effort should be taken to ensure that all employees are aware of upcoming changes, timelines, and the reasons behind the change. This can be accomplished with service talks, postings, or newsletters in paychecks. All three options may be used to ensure that the message is heard and that employees are not confused and buy in to the programs.

In general, less effort is required for operational and process changes. These options can usually be implemented in a much quicker fashion than equipment or material changes. A general outline of the scope of an implementation effort is provided below:

- Approve the project or program.
- Finalize the specifications and design for each alternative.
- Submit and gather bid requests and quotes (if necessary).
- Complete and submit a purchase order.
- Receive and install the equipment.
- Finalize operating and maintenance procedures.
- Train affected employees.
- Start the project or program.
- Complete regulatory inspections.
- Track implemented project cost savings and quality improvements.

The improvement phase is designed to identify actions remedying the root cause of the waste or inefficiency of a process [2]. The following tools can be implemented to solve a problem:

- Systems thinking
- Testing of hypothesis

- Comparative experiments
- Design of experiments

Systems thinking involves making a decision based on data and facts. Often decisions are based on thoughts and studies are not done to find the true causes of problems. Systems thinking is conducting tests and using tools to find these problems and fix the problems based on these facts.

A statistical hypothesis is an assertion or conjecture concerning one or more populations. Testing of hypothesis is an inferential technique used to make a statement about an activity or process based on its output [2]. One can either fail to reject a hypothesis or reject it. Fail to reject implies that the data do not give sufficient evidence; on the other hand, rejection implies the sample evidence refutes the hypothesis. Rejecting the hypothesis means when the hypothesis is true there is a small probability of that sample occurring again. There are always two hypotheses: the null and the alternative. The test is against the null; if the test is rejected then the alternative is accepted.

Comparative experiments is measuring a control group then conducting an experiment and seeing if there is a difference in results [2]. The design of experiments relates to comparative experiments. It is an unbiased way of creating an experiment while getting the wanted answers to the problem. The three goals of an experiment are

1. To verify a hypothesis
2. To refute a hypothesis
3. To validate a hypothesis

The experimenter should clearly set forth his or her objectives before proceeding with the experiment. The experiment should be described in detail. The treatments should be clearly defined.

8.3.10 Step 10: Validate Program versus Goals

Many companies require a validation process to ensure that projects and programs have met the goals that were set at the beginning of the project. This process aids in future planning by tracking the estimated versus actual results. This includes validating that the project or program was installed at or below cost, that it is operating within the expense and revenue limits, and that it is achieving the quality improvement and reduction goals. Even if an organization does not require a validation process, it can be a very valuable tool for future planning processes to identify where estimation errors occurred and take effort to correct them. Alternatives that do not meet the established goals or expected performance expectations may require rework or modifications. It is also critical to store warranties and contracts from vendors prior to the installation of the equipment. Also, the experience gained in implementing an option at one facility can be used to reduce the problems and costs of implementing options at subsequent facilities.

An actual performance analysis should be completed for each piece of equipment, process, or material change that was implemented. The analysis provides a standardized method to compare project performance against estimates in terms of:

- Project duration
- Implementation cost
- Operating expenses and revenue
- Cycle time and productivity
- Product or process quality
- Safety

It is useful to emphasize that the purpose of the worksheet is not a "gotcha" game, but a method to improve future project estimates and learn from mistakes if applicable. In terms of project duration, the alternatives should be evaluated based on the time required to implement the alternative versus the original estimate. Explanations should be provided for large deviations, such as "an additional two weeks required to obtain building permits." Actual implementation cost should be analyzed versus estimates, in addition to operating expenses (including labor, materials, and utilities), revenue generation (increased sales from higher customer satisfaction), and cost savings from process changes. For example, if the purchase of a new punch press was expected to increase throughput by 15% and the punch press only increased throughput by 8%, a root cause analysis should be conducted to explore and improve the deviation. Any cycle time or productivity deviations from the original estimates should also be explored. These deviations could have a very negative effect on the organization's profitability and in most cases are very closely watched by upper management. The same goes for product quality. Finally, any safety concerns should be addressed immediately. A walkthough by the team leader, safety captain, and area supervisor can quickly identify and resolve these issues. To aid in the validation process a worksheet is provided in Figure 8.34. The validation process should be performed within four to eight weeks of implementation.

8.3.11 Step 11: Monitor and Continually Improve Performance

After the Lean Six Sigma project or program has been implemented and validated, it must be monitored on a periodic basis to ensure that it is still performing as planned. If needed, the team should also make any necessary adjustments. This includes monitoring the quality, operational, and financial performance versus the goals. In addition, emphasis should be placed

Lean Six Sigma Validation Worksheet

Company Name ______________________
Location: ______________________
Date: ____________
Alterative Description: ______________________
Alternative Tracking Number: ___

	Estimated	Actual	% Difference	Comments
Project Duration				
Space Requirements				
Implementation Cost				
Machine Costs				
Machine cost				
Site development				
Material Costs				
Building modification costs				
Permitting and inspection costs				
Contractor fees				
Start up costs				
Initial training costs				
Operating Expenses				
Utility cost impacts				
Input material changes				
Labor cost impacts				
Supervision cost impacts				
Maintenance cost impacts				
Operating and maintenance supply impacts				
Changes in Overhead costs				
Operating Savings and Revenue				
Reduced operating costs				
Revenues from increased productivity				
Revenues from the sale of end products				
Cycle Time				
Product Quality				
Safety				

FIGURE 8.34
Lean Six Sigma validation worksheet.

on continuous improvement to enhance current Lean Six Sigma programs and to identify new opportunities. It may be beneficial to conduct period reviews, facility walkthroughs, or employee interviews by the original Lean Six Sigma team to accomplish these goals. When evaluating the program it is important to:

- Keep track of program success and to build on past successes.
- Identify new ideas for quality improvement and cost reduction.
- Identify areas needing improvement.
- Confirm document compliance with state or local regulations.
- Determine the effect of new additions to the facility or program.
- Keep employees informed and motivated.

New product or process changes should also be evaluated at the onset to ensure that the design is consistent with the organization's Lean Six Sigma goals. This is easily accomplished by adding "Lean Six Sigma review" to the new product or process checklist or standard operating procedures.

The Lean Six Sigma program should be viewed as a continuing effort versus a one-time project. Generally, the first Lean Six Sigma project and the resulting alternatives that are implemented will target only the high-priority or high-impact areas. Once these high-priority areas have been addressed, the team can focus on lower-priority areas. From a systems standpoint, the ultimate goal of the team is to optimize all quality, operational, and costs for the organization. The frequency with which the additional Six Sigma projects are conducted will depend on the budget of the company. In general, organizations that conduct assessments one to four times per year have achieved paybacks. In addition, if there are special circumstances that indicate the need for further review, a Lean Six Sigma assessment should be conducted; these special circumstances include:

- Major shift that lowers customer satisfaction
- A change in raw material or product requirements
- Higher operational or supplier costs
- New regulations
- New technology
- A major event with undesirable consequences (such as increased competition)

To be truly effective, an organizational culture of Lean Six Sigma must be fostered within the organization. Executive management must ensure this through repeated communications and acknowledgments for success stories from individuals or business units. This will make Lean Six Sigma an integral part of the organization's operations.

The control phase is designed to maintain the benefits of the improved process. Tools that are commonly used throughout this process are:

- Control charts
- Documentation
- Change management
- Communication
- Reward and recognition
- Check sheets

Control charts are used to identify variation whether it is random or assignable [2]. They record a task over a given amount of time and are plotted against the specification limits to observe if the process is in control. Documentation is the recording of the new process steps to provide standards and guidelines to the employees [2]. An example of documentation is ISO 9000. Communication is essential throughout the process but to implement something new and to control it is vital to the success of the new process [2]. Rewarding and providing recognition to all involved is also important, giving everyone a sense of pride in the hard work put into the project. Check sheets can be used to show where different problems could arise, and can keep track of different steps throughout the process or times that different products arrive. Check sheets are a good way to monitor a process and can be used for various aspects of tasks.

References

1. Pande, P.S., Neuman, R.P., and Cavanagh, R.R. (2000).*The Six Sigma Way*. New York: McGraw-Hill.
2. Trusko, B.E., Pexton, C., Harrington, J.H., and Gupta, P. (2007). *Improving Healthcare Quality and Cost with Six Sigma*. Trenton, NJ: Pearson Education.
3. iSix Sigma. (2014). *Definition of Cycle Time*. http://www.isixsigma.com/dictionary/cycle-time/
4. Overall Equipment Effectiveness Definition. (2014). http://OEE.com
5. Tompkins, J.A., White, J.A., Bozer, Y.A., and Tanchoco, J.M.A. (2010). *Facilities Planning*. New York: Wiley.
6. Pande, P. and Holpp, L. (2002). *What Is Six Sigma*. New York: McGraw-Hill.
7. Yang, K. Design for Six Sigma and value creation. (2005). *International Journal of Six Sigma and Competitive Advantage*, 1(4): 355–368.
8. Kwak, Y.H. and Anbari, F.T. (2006). *Benefits, Obstacles, and Future of Six Sigma Approach*. New York: Elsevier Science.
9. ASQ. (2014). *FMEA*. http://asq.org/learn-about-quality/process-analysis-tools/overview/fmea.html
10. iSigma. (2014). *Sigma Level*. http://www.isixsigma.com/dictionary/sigma-level/

9

Training and Implementation

9.1 Introduction

Training is a critical element of any successful program launch. Training teaches all affected employees the "how" of each process change. This is in contrast to an education program that teaches the "why" behind a process change. Both are very necessary elements, but the primary purpose of training is related to execution of the selected strategies. The training should be focused on the employees who will be performing the new process related to the Lean Six Sigma initiatives. The training should focus on an understanding of the basic principles behind the Lean Six Sigma strategies and processes; employees should understand the integration of these tools with current techniques. A just-in-time training approach is usually most effective so that employees will be applying the training very shortly after learning the concepts. The delivery method for the training should also be taken into consideration. Poor, inadequate, or untimely delivery of training material can make the best material into a less than effective exercise that fails to teach employees the needed skills regarding the process changes. The goal of effective training is related to the employee's ability to perform new tasks effectively on the job.

9.2 Training Strategy

The key strategy to training involves organizing and delivering material in such a way that trainees can immediately apply the concepts. The training will be fresh in the employees' minds and their level of excitement with the Lean Six Sigma project will be at its highest. On-the-job training is highly recommended as it teaches practical application-based skills that employees will use. For example, if a new punch press is installed, the operators should be given a detailed training session on the machine's operation, a standard operating procedure, a safety talk, and a list of contacts in the event the machine malfunctions or if they have improvement ideas. One-on-one coaching should also be made available as needed.

During the training session, example problems and trainee exercises may be beneficial to allow the trainees to work together to solve Lean Six Sigma problems in their work units and identify new cost and quality opportunities. A group discussion can also be a useful component of the training exercise to gather feedback and employee perceptions. The trainees should be made aware of the key metrics of the Lean Six Sigma program. Specifically, the environmental and economic impacts in their work units should be presented and discussed. By discussing the financial and quality benefits of Lean Six Sigma to the employees, they can be made aware that the programs will enhance job security by improving the bottom line of the organization. The tracking and monitoring methods of the waste reduction programs should also be discussed. Trend charts that display quality and economic performance versus expectations should be presented as they reinforce and highlight the continued benefits of the Lean Six Sigma programs. The location of these charts and how to read them should also be provided. Suitable locations include common areas such as lunchrooms or informational boards and should be in line with organizational policies. Figure 9.1 provides an example performance chart that may be used as a template for these trend charts.

In conjunction with the training session, the Lean Six Sigma and management team should carefully consider a certification or recognition process. Most trainees will be more motivated if there is some type of reward process from a simple thank you from the unit supervisor, a catered lunch, door prizes, or monetary prizes for meeting established goals. Employees may gain a stronger sense of teamwork if a certification process is involved

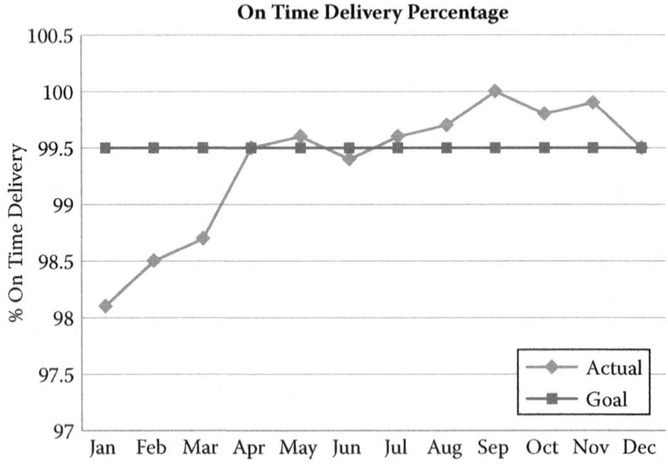

FIGURE 9.1
Sample performance tracking chart.

with new equipment deployment. For example, if a new punch press is installed, the work unit can be provided with a "certification checklist." If all items on the checklist are met, the process can be officially certified and the work unit can be recognized or presented with an award mentioned above. Common elements of the checklist may include:

- Adherence to safety standards
- Adherence to the standard operating procedure
- Housekeeping and cleanliness
- Achievement of quality goals over a three-month timeframe
- Achievement of cost reduction/revenue goals over a three-month timeframe

Many organizations establish internal Six Sigma green belt and black belt certifications. These certifications typically involve formal training related to statistics and the Lean Six Sigma tools in addition to hands-on Lean Six Sigma project work. Many organizations that have internal Six Sigma green and black belt certification require a written test in addition to an assessment by a Lean Six Sigma expert or leader within the organization.

9.3 Training Agendas

A typical training agenda includes a one-day workshop for the employees that will apply the new Lean Six Sigma processes. These employees may include production workers, maintenance, management, custodial, and clerical. The training should discuss the new process changes, equipment, the goals of the program, tracking methods, continuous improvement for the Lean Six Sigma program, and the need for a quality mindset to change the organization culture. Below is a list of "lessons learned" that may be useful when developing agendas:

- Ensure that the measurement and tracking system is unbiased, accurate, and updated in a regular and timely manner.
- Regularly communicate results of the Lean Six Sigma program to all employees via newsletters, service talks, and bulletin board postings.
- Ensure a feedback mechanism to strengthen the program and foster employee buy-in.
- Create a reward program for meeting the program goals (some organizations prefer time off, bonus checks, luncheons, or even a donation made in the organization's name to a charity organization).

Time	Agenda Item
9:00 a.m. to 9:30 a.m.	Registration and Continental Breakfast
9:30 a.m. to 9:45 a.m.	Welcoming Address by CEO
9:45 a.m. to 10:15 a.m.	Lean Six Sigma Program Overview and Goals
10:15 a.m. to 11:30 a.m.	Overview of Program Changes and New Procedures
11:30 a.m. to 12:30 p.m.	Catered Lunch
12:30 p.m. to 1:00 p.m.	Tracking Reports and Monitoring Processes
1:00 p.m. to 1:30 p.m.	Feedback and Continuous Improvement
1:30 p.m. to 2:00 p.m.	Rewards Program
2:00 p.m. to 3:30 p.m.	Facility Tour and On-the-Job Training for New Processes
3:30 p.m. to 4:00 p.m.	Question Session and Closing Remarks

FIGURE 9.2
Sample training agenda.

A training agenda should be developed prior to the training and disseminated to the employees who will attend the training session. The agenda is a useful tool to prepare employees for the training by setting expectations, time requirements, and desired outcomes from the training. If an exam or on-the-job application of the training will be provided, it should be stated in the agenda. The dress code for the training should also be stated, inasmuch as activities may involve upper management, outdoor activities, or shop floor activities. Figure 9.2 provides a sample agenda that may be used as a template.

A separate day-training session/workshop should be held for executives and managers to gain a better understanding of the program and the facilitation process. The program should emphasize the manager's role in coaching employees and the execution and monitoring of the system.

10

Common Pitfalls

10.1 Introduction

Although Lean Six Sigma can lead to tremendous positive results for organizations, if the process is not implemented properly it can lead to negative consequences as well. This chapter discusses several of the common pitfalls that should be understood and avoided before an organization implements Lean Six Sigma. The pitfalls listed in the following sections are presented in the order of highest impact to lowest impact; recommendations and best practices to avoid these pitfalls are additionally discussed in Chapter 11.

10.2 Common Pitfalls

The first common pitfall of Lean Six Sigma is the belief that it is universally applicable to all organizations [1]. Although the general philosophies and methods apply to most organizations, the implementation and specific methods must be selected and modified based on the specific resources and constraints of the organization. Some organizations have certain technical skill sets, engineers, and quality personnel that are better suited for Lean Six Sigma implementation. This leads into the second common pitfall, that professional outside assistance from statisticians or consultants is not needed [1]. Outside experts can bring new perspectives, tools, and guidance to the Lean Six Sigma implementation process, especially for organizations implementing Lean Six Sigma for the first time. A third common pitfall is a bigoted "in data we trust" mentality that leads some organizations to go overboard with statistical evidence rather than experience or insights of business leaders [1]. The fact that Lean Six Sigma is data driven is generally a strength, but if it is strictly enforced 100% of the time, it may exclude very valuable insights from proven business leaders that lack strong statistical evidence. Outside the common pitfalls, other "smaller" pitfalls also exist and should be taken into account when considering the implementation of Lean Six Sigma; these other pitfalls include:

- Implementing too fast
- Not providing enough resources in terms of talent or employee time
- Lack of training and experience
- Not solving the correct problem
- Not involving and informing all employees of the process
- Not properly tracking results

Deployment mistakes can also derail a well-developed Lean Six Sigma strategy. A. Chandra identified the following five Six Sigma deployment mistakes [2]:

- Leadership indifference
- Faulty deployment strategy
- Too much stress on training and certification
- Incorrect problem selection
- Segregating the effort

In terms of Lean manufacturing, Ulises Pabon identified the 15 most common mistakes in Lean implementation [3]:

- Thinking of 5S as something you do to an area
- Imposing 5S top-down, with limited involvement bottom-up
- Equating waste reduction with cost cutting
- Remaining aloof to the larger global end-to-end value stream
- Assuming your future state VSM is nothing more than your current state VSM with the identified improvement opportunities corrected or addressed
- Equating visual workplace with top-down visual communication
- Viewing TPM as an improvement initiative that exclusively relates to engineering and maintenance personnel
- Using OEE to evaluate operations rather than as an improvement gauge
- Equating standard work with procedures.
- Engaging in "industrial tourism" and thinking you are benchmarking
- Pursuing a one-size-fits-all solution to production planning and control
- Forgetting to reduce supermarket inventories once established
- Preconditioning continuous flow to waste elimination

- Believing you will achieve a Lean transformation by applying Lean tools.
- Betting your strategy on Lean

Chapter 11 provides best practices to overcome these common pitfalls.

References

1. Goh, T.N. (2010). Six triumphs and six tragedies of Six Sigma. *Quality Engineering*, 22:299–305.
2. Chandra, A. (2014). *5 Six Sigma Deployment Mistakes—And How To Avoid Them*. http://www.isixsigma.com/new-to-six-sigma/getting-started/5-six-sigma-deployment-mistakes-and-how-avoid-them/ Retrieved Dec. 16, 2014.
3. Pabon, U. (2010).*The 15 Most Common Mistakes in Lean Implementations*. San Juan, Puerto Rico: QBS, Inc.

11

Best Practices and Lessons Learned

11.1 Introduction

As discussed in Chapter 10, several common pitfalls exist related to implementing Lean Six Sigma. Fortunately, many best practices and lessons learned are available and this chapter discusses many of the most common.

11.2 Best Practices and Lessons Learned

GE Corporate Research identified 20 key lessons related to implementing Six Sigma based on their 20-year history of success in the area; the lessons are presented below in a logical sequence not meant to reflect importance (1).

- The time is right.
- The enthusiastic commitment of top management is essential.
- Develop an infrastructure.
- Commit top people.
- Invest in relevant hands-on training.
- Select initial projects to build credibility rapidly.
- Make it all pervasive, and involve everybody.
- Emphasize design for Six Sigma (DFSS).
- Do not forget design for reliability.
- Focus on the entire system.
- Emphasize customer critical to quality characteristics (CTQs).
- Include commercial quality improvement.
- Recognize all savings.
- Customize to meet business needs.

- Consider the variability as well as the mean.
- Plan to get the right data.
- Beware of dogmatism.
- Avoid nonessential bureaucracy.
- Keep the toolbox vital.
- Expect Six Sigma to become a more silent partner.

In addition to the above list of lessons of best practices identified by GE, several best practices also exist that should be considered when implementing Lean Six Sigma. The lessons learned listed below are presented in the order of highest impact to lowest impact.

The first best practice of Lean Six Sigma is the use of a common realistic metric for quality assessment and improvement [2]. DPMO and Sigma level are the commonly accepted metrics that are universally applied in most Six Sigma projects. These metrics are also easily understood by upper management and do not require high-level mathematics and statistics to explain, understand, and benchmark to other programs or companies. The second best practice of Lean Six Sigma is clear assignment of roles and responsibilities in performance improvement efforts [2]. It is critical to define the Lean Six Sigma team and member responsibilities. This allows for better tracking of results and accountability. The third best practice of Lean Six Sigma is to allow capabilities to grow for larger roles for business competitiveness; unlike many other quality tools or certification systems that remain essentially the same throughout their useful life, Six Sigma is organic [2].

References

1. Hahn, G.J. (2005). Six Sigma: 20 key lessons learned. *Quality and Reliability Engineering International*. 21:225–233.
2. Goh, T.N. (2010). Six triumphs and six tragedies of Six Sigma. *Quality Engineering*, 22:299–305.

Section III

Lean Six Sigma Implementation via Case Studies

12

Case Study 1: Logistics and Transportation Networks

12.1 Introduction

This case study describes a Lean Six Sigma project at the US Postal Service involving the use of Six Sigma tools to enhance customer service by improving on-time delivery at an airmail processing center (AMC). Specifically, the largest contributors to delivery failures were examined and reduced at the Columbus, Ohio AMC. The primary metric for this project was the percentage of letters that were processed at the Columbus AMC and were delivered on time based on the postal standard of three days.

Annually, the Columbus AMC processes and distributes over 50 million first-class letters generated in central Ohio that are transported by air. These letters are transported utilizing up to 175 daily airline flights from the Port Columbus International Airport and the Rickenbacker Airport; both airports are located within 15 miles of the city of Columbus. When the analysis began, 8.7% of all letters processed at the Columbus AMC were not delivered on time based on the organizational goal.

This remainder of this case study provides an overview of US Postal Service operations and on-time delivery measurement (which is a key indicator for customer service), a discussion of the Six Sigma methodology applied, and a discussion of the results of the project.

12.2 Background

12.2.1 About the US Postal Service

The US Postal Service delivers more than 213 billion letters and packages annually, operates nearly 37,000 post offices, stations, and branches and delivers to more than 146 million delivery points. This vast network requires a highly developed distribution system to move the mail efficiently and effectively by air and ground. Transportation expenses for 2006 were $6.0 billion; of that amount 45.8%, or $2.8 billion was related to air transportation. The US Postal Service operates 77 airmail processing centers across the country to process and distribute letters and packages via air transportation. The US Postal Service does not own a fleet of planes; instead contracts are made between commercial airline and private corporations to provide the air transportation.

This study took place at an airmail processing center located in Columbus, Ohio. The Columbus Airmail Processing Center employs 72 workers responsible for processing and distributing letters generated in the central Ohio area that will be transported by air.

12.2.2 Service Measurement

Outstanding customer service is the top priority of the US Postal Service. On-time delivery for first-class mail is independently measured by IBM Consulting Services from collection box to mailbox. This includes first class with overnight, two-day, and three-day delivery commitments based on distance. Most mail that is transported via airplane has a three-day service commitment; this is the primary type of mail processed at the Columbus AMC and is the focus of this study.

12.2.3 Airmail Processing Overview

The Columbus AMC processes over 50 million letters annually, utilizing a combination of automated and manual processes. When a handling unit (which includes letter trays, sacks, and large parcels) arrives at the AMC, it is placed onto a conveyor belt that feeds it into a scanning machine. This scanning machine reads a barcode that represents the five-digit ZIP code that the handling unit is to be sent to for delivery. A database called the Surface Air Management System (SAMS) is used at each AMC to assign handling units to appropriate air transportation and execute contingency rules using planned and active airline route information. Based on the handling unit's destination as determined by the ZIP code, it is assigned to a flight in SAMS. Once an available flight is determined, the handling unit is assigned to the specific flight and a dispatch and routing tag is generated and attached to the handling unit. The dispatch and routing tag essentially serves as an airline ticket for the handling unit.

These finalized handling units are then staged in a loading area and are collected by the corresponding ground handlers for each airline. The ground handlers then complete a final scan on each item as it is loaded on the assigned flight. These handling units are then flown to their destinations, offloaded by ground handlers at the arrival airport, and transported to the receiving AMC to be processed for delivery. Figure 12.1 provides a snapshot of US Postal Service operations in terms of size and scope.

Columbus Airmail Processing Center at a Glance
• Built in 1982
• 85,000 square feet
• 72 employees on three shifts daily and weekends
• Processes 50 million pieces of mail annually
• One of 77 Airmail Centers in the US

FIGURE 12.1
Snapshot of US postal operations.

12.3 Application of Lean DMAIC Process

12.3.1 Define

The objective of this project was to improve the on-time delivery of mail generated in central Ohio and processed at the Columbus AMC to a rate equal to or above the organizational goal of 90%. The primary metric for this project was the percentage of letters that were processed at the Columbus AMC and were delivered on time based on the postal standard of three days as calculated independently by IBM Consulting Services. IBM Consulting Services' sampling method is the primary tool to measure on-time delivery for the postal service. The process was baselined at an on-time delivery rate of 81.3% (8.7% short of the goal and 18.7% short of zero defects) as determined independently by IBM Consulting Services for the preceding six months. The Six Sigma team was comprised of the local US Postal Service management team, front-line supervisors, mail handlers, an industrial engineer, airline management, and two airline ground handlers.

12.3.2 Measure

In this phase of the project the key metrics were identified, the data collection process was developed and executed, the performance and variation of the current system was base-lined, and the initial Sigma level was calculated. Included in this phase was the development of a SIPOC diagram and process map. Figure 12.2 is the SIPOC diagram and Figure 12.3 is the process flowchart that were created as part of the measure phase to describe and understand the process.

The output variable analyzed for this process was the time required from the point a letter entered into the postal network (via a collection box, post office, or business mail entry unit) to the time the letter was delivered into the addressee's mail receptacle. The input variables (potential vital Xs) included:

- Incorrect addresses on letters
- Postal transportation delays

Suppliers	Inputs	Processes	Outputs	Customers
>Businesses	>Letters	See Below	>Delivery of Letter	>Businesses
>Consumers	>Letter sorting machines		>Payment	>Consumers
>Mail sorting houses	>Flight data systems			
	>Airplanes			
	>Trucks			
	>Fuel			

SIPOC Diagram

FIGURE 12.2
SIPOC diagram.

- Airline transportation delays
- Airline flight cancellations
- Flight or truck capacity issues
- Sort program errors
- Labeling issues (machines unable to read barcodes)
- Surface Air Management System database errors

Data were collected for this project over a three-month period utilizing a combination of failure analyses for late arriving mail, database integrity reviews, delay studies, sort program reviews, and label reviews. The data were collected by the cross-functional team using standardized forms.

Meaningful Six Sigma metrics were established based on the data and goal of the project. These metrics included the process cycle time, Sigma quality level, process capability, defects per million opportunities (DPMO), and yield. A quality control chart was also created with upper and lower bounds.

The US Postal Service definition of on-time delivery for first-class letters transported by airplane is delivery to the addressee within three days from when the letter entered the postal network (or 72 hours). The national goal of the postal service is that 90% of these three-day letters are delivered within 72 hours. A two-month sampling of letters originating in the Columbus area and flown around the United States was gathered with a daily sample size of 50 mail pieces. A p chart was created based on the data gathered. The fraction rejected was $p = 0.176$ with an upper control limit of $UCL = 0.338$ and a lower control limit $LCL = 0.014$ for the 50-piece sample size. In summary, 17.6% of letters sent via airmail were not arriving on time; conversely 82.4% of letters were arriving on time (7.6% short of the organizational goal of 90%). Based on the calculated control limits for the p chart, a given sample of 50 pieces could have between 0 to 8 pieces and still be in control. With a mean fraction of 17.6% of letters arriving late, the data indicate the process is not capable of achieving the 90% goal for on-time delivery.

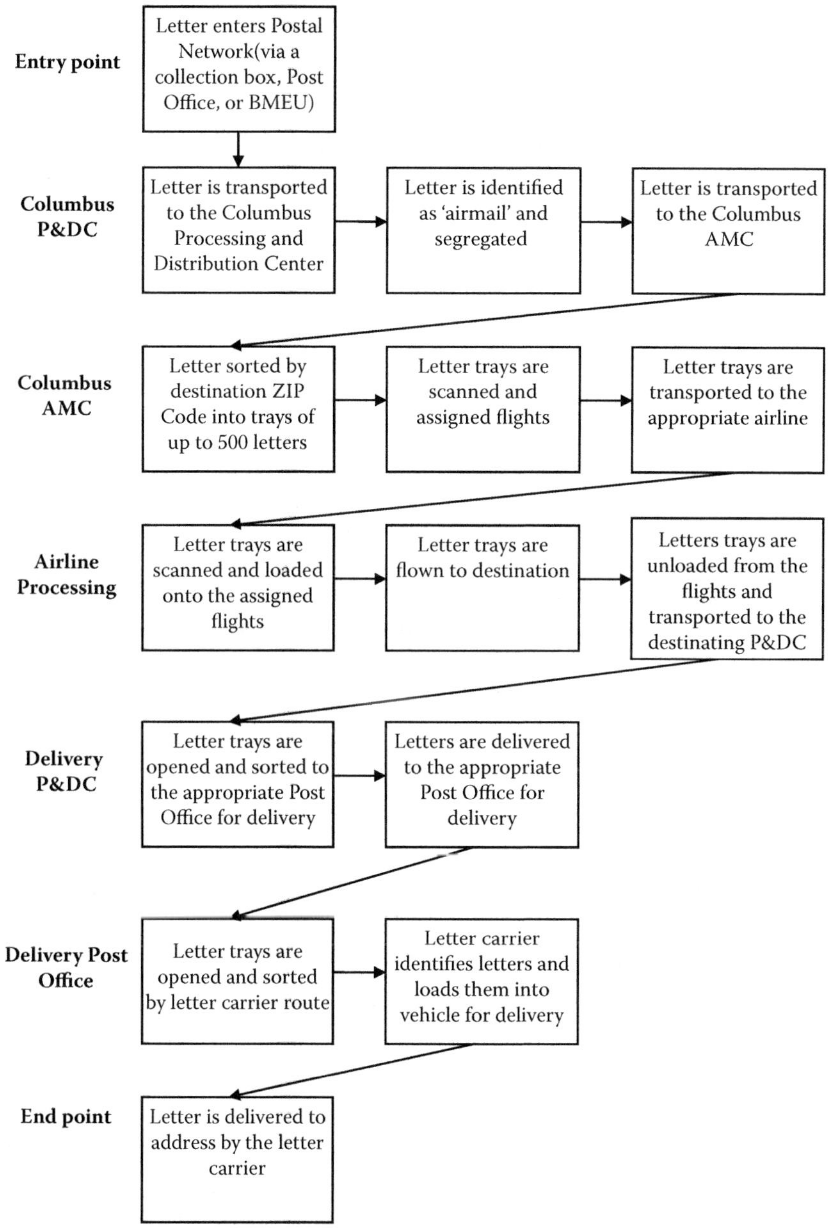

FIGURE 12.3
Process flowchart.

The 17.6% late-arriving mail rate translates into 176,000 defects per million opportunities or a 2.4 Sigma level for the current process considering the 50 million letters the Columbus AMC processes annually.

12.3.3 Analyze

Various Six Sigma tools were applied to identify the vital *X*s or the root causes of delivery failures originating in the Columbus area. The tools and data analysis methods applied included:

- Pareto analysis
- Five Why analysis
- Root cause analysis
- Independence hypothesis testing (using chi-square)
- FMEA

These tools were applied to identify the assignable causes leading to the gap between the current performance and the goal. Numerous reasons such as incorrect addresses, incorrect flight information, poor quality labels, inadequate capacity, and cancelled flights contributed to the delivery failures. After developing the process map and collecting data for the three-month timeframe, Pareto and five Why analysis indicated airline ground handler delays (26%) and SAMS data system errors (23%) were the top reasons for failures. Please see Figure 12.4.

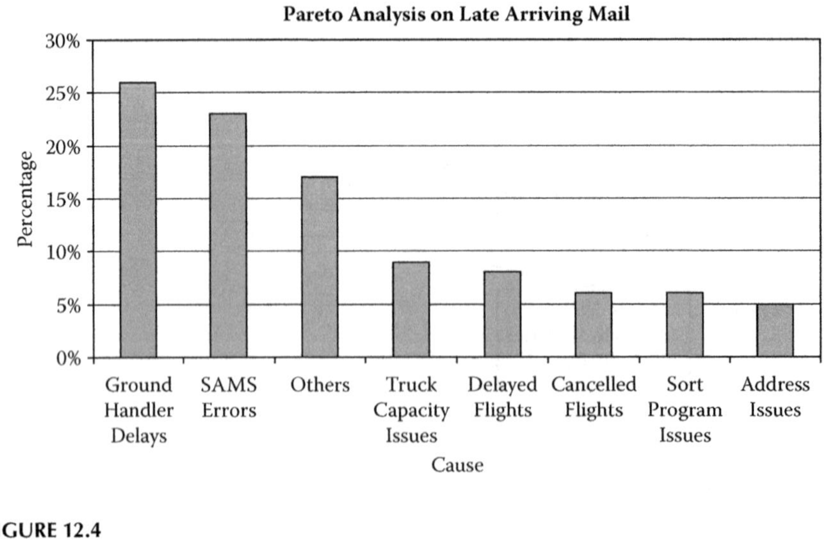

FIGURE 12.4
Pareto analysis of late arriving mail.

Using the three months of data that were collected, several hypothesis tests were developed and conducted. The purposes of the tests were to test the hypothesis of independence among variables and test whether the frequency distributions fit the expected uniform distribution. Specifically, chi-squared tests were conducted on variables including:

- Day of the week
- Location the letter was sent (segmented into geographic regions)
- Presence of known system conditions that could delay the letters (such as ground handler delays, SAMS errors, airline delays or cancellations, or sort program issues)

The following hypothesis tests were established:

H_0: The distribution of delivery failures of airmail originating in Columbus is uniform for each day of the week the mail was inducted into the system.

H_a: The distribution of delivery failures of airmail originating in Columbus is not uniform for each day of the week the mail was inducted into the system.

The tests were conducted at the $\alpha = 0.05$ level of significance with $k - 1$ degrees of freedom using the chi-squared statistic. Tables 12.1 through 12.3 are the data tables that were used for the hypothesis tests.

TABLE 12.1

Day of Week

	Monday	Tuesday	Wednesday	Thursday	Friday	Saturday	Total
On time	523	534	510	517	528	503	3,115
Late	107	96	120	113	102	127	665
Total	630	630	630	630	630	630	3,780

TABLE 12.2

Geographical Region

	Pacific Area	Western Area	Southeast Area	Southwest Area	Northeast Area	Total
On time	776	601	659	720	359	3,115
Late	174	119	151	160	61	665
Total	950	720	810	880	420	3,780

TABLE 12.3

System Conditions Present

	SAMS Errors Present	Ground Handler Delays Present	Truck Capacity Issues	Delayed Flights	Cancelled Flights	Sort Programs Issues	No Issues Present	Total
On time	385	360	320	215	83	167	1,585	3,115
Late	126	99	75	37	14	29	285	665
Total	511	459	395	252	97	196	1,870	3,780

For the day of the week hypothesis test, the null hypothesis was not rejected. At the 5% confidence level, there is enough evidence to conclude that the distribution of delivery failures by day of the week is uniform (p-value of 0.2024). In other words, the day the letter was inducted into the system does not affect on-time delivery.

For the geographical hypothesis test, the null hypothesis was not rejected. At the 5% confidence level, there is enough evidence to conclude that the distribution of delivery failures by geographical region is uniform (p-value of 0.3475).

For the presence of system conditions that could affect on-time delivery the null hypothesis was rejected. At the 5% significance level, there is enough evidence to conclude that the presence of known system conditions does indeed affect on-time delivery (p-value of 0.000009). Taking a closer look at the presence of system conditions, the on-time and late percentages were calculated as displayed in Table 12.4.

Based on the percentages in Table 12.4, SAMS errors present and ground handler delays present significantly lowered on-time percentages versus the other system conditions. Both conditions resulted in a 4% to 7% decrease from the mean of 82.4% on time. Based on the hypothesis test findings and upon further analysis, a failure mode effects analysis (FMEA) was conducted. The Six Sigma team analyzed the data and system to determine the failure modes, potential effects, severity, occurrences, and detection. The key findings from the study are listed below and the actions taken are listed in the Improve section of this case study.

TABLE 12.4

On-Time Delivery Percentages by System Conditions

	SAMS Errors Present	Ground Handler Delays Present	Truck Capacity Issues	Delayed Flights	Cancelled Flights	Sort Programs Issues	No Issues Present	Total
On time	75.3%	78.4%	81.0%	85.3%	85.6%	85.2%	84.8%	3,115
Late	24.7%	21.6%	19.0%	14.7%	14.4%	14.8%	15.2%	665
Total	511	459	395	252	97	196	1,870	3,780

- *Ground handler delays:* By shadowing the commercial airline ground handling crews that load mail onto departing flights, the team found that several delay points existed in the transportation process from the AMC to the airplane. The goal is to transport the mail from the facility within 30 minutes of arrival at the airline staging area. The airlines that serve the Port Columbus International Airport achieved the 30-minute goal for 69% of departing flights.
- *SAMS database issues:* The team identified incorrect information and lack of real-time monitoring of the Surface Air Management System. SAMS displays active routes and routes that have closed out in the last 12 hours. Several situations were identified where handling units were being assigned to closed flights (flights that had met capacity), cancelled flights, or flights that would not provide adequate time to transport the handling units to the plane for loading before takeoff. Delayed flights also were identified as a vital *X* as many of these flights would deliver the handling units to the arrival processing center dead on arrival to meet the delivery goal. From a three-day survey of SAMS, 11% (50 out of 455) of the flights had incorrect information in terms of capacity, availability, or time of actual departure.

12.3.4 Improve the Process

After identifying the potential vital *X*s, the hypothesis was tested and validated as discussed in the previous section. Meetings were held with the ground handler crews for each of the major airlines and a tracking system was developed to record transfer times to the AMC after flight touchdown. Results were shared daily with the airline's responsible managers. After meeting with all stakeholders, a standardized process was created and documented. The goal of the standardized process was to "mistake-proof" any preventable delivery failures. Within four weeks of tracking delays, 98% of arriving flights were loaded and transported to the processing facility within 30 minutes of the plane takeoff.

An additional two months of data were collected and an additional hypothesis test was developed and conducted to validate the improvements. The purpose of the test was to examine the hypothesis of independence among variables and test whether the frequency distributions fit the expected uniform distribution. In other words, if the new distribution after the changes were implemented fit the normal distribution, there was sufficient evidence to conclude the two primary assignable causes (SAMS system errors and ground handler delays) were removed from the system. Specifically, a chi-squared test was conducted on variables for the presence of system conditions that could affect on-time delivery; the null hypothesis was rejected. At the 5% significance level, there is enough evidence to conclude that the presence of known system conditions no longer affect on-time delivery

TABLE 12.5

Summary of Validation Test

	SAMS Errors Present	Ground Handler Delays Present	Truck Capacity Issues	Delayed Flights	Cancelled Flights	Sort Programs Issues	No Issues Present	Total
On time	189	165	218	170	96	145	1,426	2409
Late	13	11	9	6	5	6	61	111
Total	202	176	227	176	101	151	1,487	2,520
On time	93.6%	93.8%	96.0%	96.6%	95.0%	96.0%	95.9%	3,115
Late	6.4%	6.3%	4.0%	3.4%	5.0%	4.0%	4.1%	665

TABLE 12.6

Combined Test Results

Delivery Failures	Baseline	Test Results
Defect rate	17.6%	4.4%
DPMO	176,000	44,000
COPQ		$24,500
Total annual savings		$15,000
Sigma level	2.4	3.2

(p-value of 0.1686). Table 12.5 is a summary of the data collected for the validation test.

Training courses and standard operating procedures were developed for updating SAMS before peak demand periods. Before this point, updates were automatically downloaded from a centralized office that is located out of state. As a result, a three-day survey SAMS had correct information for 99% of flights.

Table 12.6 displays the combined test results. As a result of these two process improvements, delivery failures were reduced by 75% from 176,000 to 44,000 defects per million [(176,000 – 44,000)/176,000]. This reflects an annual savings of $15,000 from reduced double handling. Based on the p-charts developed before and after the Six Sigma improvements, the process mean of late deliveries moved from a mean (center line) of 0.176 (17.6% late deliveries) to a mean of 0.044 (4.4% late deliveries). The new process mean was 5.6% above the organizational goal of 90% on-time deliveries.

12.3.5 Control the Process

After establishing the process as in control and capable of meeting the organizational on-time delivery goal of 90%, methods were put in place to ensure the process remained stable. Specifically, the following control and monitoring tools were implemented:

- A *p*-quality control chart to monitor delivery failures
- Monitoring processes for ground crew delays and SAMS database updates
- Standard operating procedures with written documentation
- Visual factory concepts including signage

To monitor "control" status, the *p*-chart discussed in the Analyze section was utilized and set to the new control limits. The fraction rejected was $p = 0.044$ with an upper control limit of $UCL = 0.131$ and a lower control limit $LCL = 0.000$ for the 50-piece sample size. In summary, 4.4% of letters sent via airmail were not arriving on time; conversely 95.6% of letters were arriving on time. Based on the calculated control limits for the *p*-chart, a given sample of 50 pieces could have between 0 to 6 pieces and still be in control. With a mean fraction of 4.4% of letters arriving late, the data indicate that the process is in control and mail is being delivered on time. In addition, ground crew delays and SAMS updates were recorded and monitored hourly, which gives the ability to monitor the defect rate continuously. Written standard operating procedures and training programs for new employees were also created and implemented to enforce and reinforce the process changes. These documents were created by a cross-functional team of airline personnel, mail handlers, and management. Signage on the workroom floor also aided in reducing common mistakes and gave simplified instructions regarding data collection and use of the *p*-chart.

12.4 Lessons Learned

In terms of this project, several key lessons were learned by the team in this particular situation. First, most of the team had very limited statistical skills and knowledge of Six Sigma. Initially, two weeks of classroom-based training sessions regarding the Six Sigma process and road map were given. It became evident this was not enough, and additional one-on-one training for each team member was given. Next, coordination efforts between the postal service and commercial airlines were somewhat challenging as well. Airline security guidelines made data collection a bit problematic for the team, but with proper planning and scheduling these issues were minimized.

This Six Sigma project demonstrated the significant benefits of utilizing data and statistical analyses to drive the team building, decision-making process, and organizational improvement in a service-oriented environment. Monitoring handling crew delays and properly maintaining the SAMS database will enable over 715,000 additional letters per year to arrive within the three-day delivery timeframe. The Six Sigma improvement project resulted in an annual cost savings of $15,000. This will increase customer satisfaction as well as reduce the operating costs associated with double handlings. This project may represent the tip of the iceberg for the US Postal Service in terms of using Six Sigma to achieve breakthrough performance.

To sustain the success of this project, the items discussed in the Control section were implemented and monitored daily by the employees responsible for each task. Specifically, the *p*-chart was discussed every day with the employees during the regular service and safety talks given at the beginning of each shift. Also, during the weekly management staff meetings, the current service scores, ground handler performance, and *p*-chart trends were discussed and analyzed. Six Sigma and customer-focused signage and trend charts were displayed throughout the AMC.

From an organizational standpoint, the US Postal Service is expanding its Six Sigma initiatives. Single, well-supported projects can be successful, and the company is ramping up to implement an organizationwide movement. To enhance the Six Sigma culture at the US Postal Service, the organization has implemented a training program for industrial engineers. This program recruits recent industrial engineer graduates who have up-to-date skill sets in process analysis and statistics. These recruits are then cross-trained in postal engineering and processing operations and are outplaced as field industrial engineers. Program curriculum involves the addition of a Six Sigma greenbelt project and a certification process.

References

1. Hahn, G.J. (2005). Six Sigma: 20 Key lessons learned. *Quality and Reliability Engineering International*. 21:225–233.
2. Goh, T.N. (2010). Six triumphs and six tragedies of Six Sigma. *Quality Engineering*, 22:299–305.

13

Case Study 2: Final Assembly Operations

13.1 Introduction

This study applied the define, measure, analyze, improve, and control (DMAIC) Six Sigma methodology and Lean principles [1] to redesign a final assembly work unit to increase capacity and reduce costs in a light manufacturing company located in the United States. The final assembly work unit that was redesigned for this study was responsible for 38% of the company's revenue stream and experienced various cost and process efficiency issues. These issues included an assembly-line imbalance, floating bottleneck operations, safety concerns, high percentages of non-value-added time for employees, housekeeping issues, and storage issues. In addition, the company was seeking to increase production in this work unit to attain a larger market share. To accomplish these goals, the organization applied the Lean Six Sigma approach to create a meaningful and practical road map. This study illustrates such an application including a case study and results. The case study focuses on the creation of value stream maps, line balancing, and facility redesign to identify and track key processes within the facility.

The manufacturing organization in this study initiated Lean Six Sigma to improve the financial outlook and attain a sustainable quality system with the following objectives:

- Retain and attract customers.
- Reduce manufacturing costs.
- Increase process flexibility and output based on market demand.
- Increase capacity.
- Increase workflow and information flow.
- Improve process documentation.
- Enhance safety and housekeeping.

The case study is organized into five sections to address these objectives. In this section an overview of Six Sigma analysis is provided; in Section 13.2, a background description of the case study is provided; Section 13.3 discusses the application of the Lean DMAIC approach; Section 13.4 discusses the improvement results; and Section 13.5 draws conclusions on the Lean DMAIC cost reduction initiatives.

13.2 Background

13.2.1 About the Company

The organization studied is a leading manufacturer of industrial cleaning equipment headquartered in northwest Ohio, USA. The company began operations in 1911 and conducts business in more than 60 countries with over $3.1 million in annual sales and employs 155 individuals. Their operations are based on one shift, and their hourly workers are unionized. The company has explored options to relocate its operations to less expensive regions, such as Mexico or the southern United States, where they could save money by reducing labor costs and overhead due to increased costs and economic downturns. Before moving forward with these plans, the company investigated reducing costs at its location.

13.2.2 Targeted Areas for Improvement

The process that was examined for improvement was the final assembly line for an automatic commercial floor scrubber that is available in three models and contains 416 different parts. The primary goal was to create a value stream map by applying Lean Six Sigma concepts to reduce waste and to develop a more suitable facility layout. The process studied has many forms of waste that a revised layout would reduce. As with many other Lean projects, the waste involved excess movements, redundancies, rework, errors, poor training techniques, and poor process flow, and were all within the direct focus.

Based on conversations with management and observations within the facility, the following areas of improvement were identified to reduce costs, improve workflow, and increase productivity:

- 40% improvement in space utilization
- 50% less work in process (WIP)
- 30% improvement in quality
- 20% improvement in worker productivity
- 20% increase in capacity
- 10% reduction in operating costs per unit
- 10% less scrap, with improvements in scrap control

13.3 Application of Lean DMAIC Process

13.3.1 Define

This first step of the DMAIC cycle identifies the goal, objectives, and scope of the project. In addition, the project team is formed and the timeline established. Tools used during this stage include brainstorming, critical to quality (CTQ) analysis, and scope definition. The primary objectives of this project were to increase capacity by 20% and reduce costs per unit by 10%, and the primary metrics were the annual capacity in terms of value of units sold per year and the weekly operating costs for the facility. As discussed later in the analysis phase, the process was baselined at an average annual capacity of 8,000 units for the preceding two-year period. Operating costs were also calculated for the same two-year period normalized based on production rates.

The project team consisted of the facility manager as the sponsor, the engineering manager as the project leader, the manufacturing engineer as the Six Sigma black belt, and several shop floor and maintenance personnel.

The project timeline was established at the first meeting of the team and is displayed below:

- Identify project objectives: 2 weeks
- Form Six Sigma team: 1 week
- Develop the project timeline: 1 week
- Define and clarify the project goal: 1 week
- Measure the current state of the operation: 8 weeks
- Analyze performance of the operation utilizing various Six Sigma tools: 10 weeks
- Improve the operation based on data provided in previous steps: 6 weeks
- Control the operation and develop quality/cost control mechanisms: 6 weeks
- Conclude the project, document results, and disband the Six Sigma team: 3 weeks
- Total Time: 38 weeks

The project time line was 38 weeks; the project was concluded within the 38-week timeframe.

13.3.2 Measure

In this phase of the project the key metrics were identified, the data collection process was developed and executed, and the performance and variation of

Suppliers	Inputs	Processes	Outputs	Customers
>Metal fabricators	>Stamped parts	See Process Flowchart in Figure 13.2	>Finished floor cleaning equipment	>Dealers
>Component suppliers	>Parts and subassemblies			>Institutions
				>Businesses

FIGURE 13.1
SIPOC diagram.

the current system were baselined. Included in this phase was the development of a SIPOC diagram and vital X analysis. Figure 13.1 displays the SIPOC diagram.

The output variables analyzed for this process were the manufacturing capacity and operating costs. The Six Sigma team conducted a brainstorming session to identify causes of variation in the system and potential vital Xs that caused the variation in the system. The input variables (potential vital Xs) included:

- Lack of real-time product and order data
- Inefficient facility layout
- Lack of communication
- Training and associated documentation
- Lack of standardized/efficient processes
- Lack of work standards
- Lack of control mechanisms

Data were collected for this project for the previous two years utilizing a combination record review, database review, and real-time data collection from current processes. The data were collected by the Six Sigma team using standardized forms.

Meaningful Six Sigma metrics were established based on the data and goal of the project. These metrics included the process cycle time, weekly operating costs, value-added activity analysis, and material handling flows/costs. The tools utilized for this phase were very effective and allowed for a strong quantified baseline of the current process in terms of operation process time and costs as discussed in Section 13.3.3.

13.3.3 Analyze

The third step of the DMAIC cycle involved studying the data and trends observed in the measurement phase to provide a data-driven basis for improvements. Several Six Sigma tools were applied to understand process capability and to identify the vital Xs, which are the root causes of delivery failures and cost overruns. The tools and data analysis methods applied included:

- Value stream analysis
- Pareto analysis
- Root cause analysis
- Failure modes effects analysis (FMEA)

Before identifying and analyzing root causes, a baseline for the current process performance was established and a process capability analysis was performed. On-site data were collected over an 8-week period and analyzed by the project team concurrently over a 14-week period.

The value stream mapping process included the flow of information and a diagram of the various processes of the product's path from customer to supplier. The value stream map was created by dividing the process in three phases:

1. Create the current-state map (process flowchart).
2. Conduct time studies to baseline current process metrics.
3. Identify opportunities for continuous improvement and methods to reduce non-value-added activities.

The remaining subsections provide more detail on each of these three steps.

13.3.3.1 Current-State Map (Process Flowchart)

This step involved conducting interviews with employees in the work unit and several days of observation. Data were collected to document and understand the steps of the process. From the information that was gathered, the following items were determined: customers, suppliers, steps of operations, flow, stock, transport, cycle times, number of operations by stations, uptime, changeover time, downtime per process, inventory stations, inputs, outputs, and delays.

Once the collection of data was complete, an analysis of the data was completed in order to understand the value stream. Figure 13.2 is a diagram of the current-state value stream map. A process flowchart is a tool used to map out the steps of a process. In this particular application, there are five elements (and symbols) that are used: operation (○), transportation (⇨), inspection (□), delay (D), and storage (▽).

Within the project process flowcharts were created for all assemblies. A process flowchart is a good tool to use in order to represent a process easily. It can also provide opportunities to identify value-added activity, non-value-added activity, and non-value-added waste easily.

Process Flowchart

Process Description: Breaker Panel Assembly
Company/Location: NSS
Recorded By: Ed Russel

Date: Friday, October 17, 2008

Summary		
Category	**No.**	**Dist.**
Operations	9	
Transfers	5	51.5 feet
Inspections	0	
Delays	1	
Storages	1	12.5 feet
Totals	16	64 feet

Step	Operation	Transfer	Inspection	Delay	Storage	Description of Method	Distance	Cycle Time	Quantity	Lead Time	Comments
1	○	⇨	□	D	▽	Get Face Plate	11 ft.	8 sec.	1	8 sec.	
2	○	⇨	□	D	▽	Get Decal and (7691221)	6.5 ft.	15 sec.	1.1	23 sec.	
3	○	⇨	□	D	▽	Place Decal		40 sec.	1	63 sec.	
4	○	⇨	□	D	▽	Place (7691221)		18 sec.	1	81 sec.	
5	○	⇨	□	D	▽	Tighten		8 sec.	1	89 sec.	
6	○	⇨	□	D	▽	Get (6491261) and (4890791) (×2)	13 ft.	7 sec.	1.2	96 sec.	
7	○	⇨	□	D	▽	Place (6491261) & Tighten		24 sec.	1	120 sec.	
8	○	⇨	□	D	▽	Get Screws	6.5 ft.	21 sec.	4	141 sec.	
9	○	⇨	□	D	▽	Get Bolts and the Red Cable	14.5 ft.	11 sec.	4.1	152 sec.	
10	○	⇨	□	D	▽	Prepare (4890791) (×2)		18 sec.	2	170 sec.	
11	○	⇨	□	D	▽	Place (4890791) (×2)		26 sec.	2	196 sec.	
12	○	⇨	□	D	▽	Tighten		7 sec.	4	203 sec.	
13	○	⇨	□	D	▽	Add the red cable		9 sec.	1	212 sec.	
14	○	⇨	□	D	▽	Checking Directions		16 sec.		228 sec.	
15	○	⇨	□	D	▽	Mistakes		15 sec.		243 sec.	
16	○	⇨	□	D	▽	Store	12.5 ft.	16 sec.		259 sec.	

FIGURE 13.2
Current-state process flow diagram.

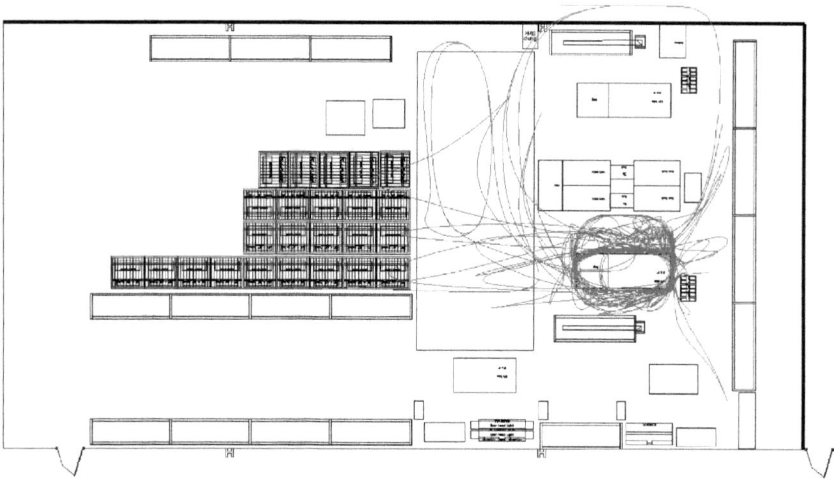

FIGURE 13.3
Current-state spaghetti diagram.

In terms of the value stream mapping, a spaghetti flow diagram was created by taking the existing layout of the facility and generating flow lines of product through the work unit. The flow diagram displays the path of an employee working on the main assembly. As displayed in Figure 13.3, the main assembly area is heavily circled as shown by the multiple lines, which indicate employee movement within the work area. There are also several lengthy trips that add up to a large amount of waste within the process. The information that was received from the spaghetti diagram helped to determine which areas were critical, which areas needed to be closer, and which areas did not need to be moved.

13.3.3.2 Baseline Time Studies of Current Process

A time study was performed on the work unit to baseline the current process metrics in terms of mean process times and the related standard deviations. This information was required to balance the assembly line process for the layout redesign. A time study is conducted by following an operator through a process or assembly and documenting the durations of the basic elements of the process. Several key concerns are crucial to an effective and accurate time study, such as pace rating, several observations of the process with the same operator and same procedure, and some degree of consistency. The information provided from time studies and the process flow sheets will be an integral part of creating a new, more efficient layout and procedure.

For the time studies six observations of each assembly were recorded. Soon after the start of taking time studies it was evident that the various operators assembled the parts differently, and their procedure changed between cycles

Element	Observations							
	1	2	3	4	5	6	Mean	St. Dev.
Get Face Plate	8	3	4	5	3	6	4.8	1.9
Get Decal and (7691221)	15	5	5	2	6	5	6.3	4.5
Place Decal	40	31	30	30	32	31	32.3	3.8
Place (7691221)	18	12	8	18	17	17	15.0	4.1
Tighten	8	6	4	7	7	6	6.3	1.4
Get (6491261) and (4890791) (×2)	7	6	6	6	6	6	6.2	0.4
Place (6491261) & Tighten	24	24	26	24	25	25	24.7	0.8
Get Screws	21	12	10	18	13	11	14.2	4.4
Get Bolts and the Red Cable	11	10	12	11	11	11	11.0	0.6
Prepare (4890791) (×2)	18	21	20	19	20	19	19.5	1.0
Place (4890791) (×2)	26	11	23	14	17	21	18.7	5.7
Tighten	7	18	7	8	7	12	9.8	4.4
Add the red cable	9	8	7	7	7	8	7.7	0.8
Checking Directions	16	45	21	12	35	36	27.5	13.0
Mistakes	15	26	0	0	21	9	11.8	10.8
Store	16	10	13	15	13	15	13.7	2.2
Total	259	248	196	196	240	238	229.5	27.0

FIGURE 13.4
Time study results.

(they did not follow a standard process). To remedy this issue, the team consulted with one trained operator to complete six rotations following the same process. The results from the time study are displayed in Figure 13.4.

13.3.3.3 Opportunity Identification for Continuous Improvement and Identification of Methods to Reduce Non-Value-Added Activities

The team identified opportunities for continuous improvement by highlighting some of the areas of waste within the processes and recommended methods to eliminate these activities; these are discussed in Section 13.3.4 related to improvements.

From these five previously mentioned steps, a Pareto analysis was performed to determine the largest contributors to cost overruns. The analysis indicated that over 50% could be contributed to a lack of standard procedures and related monitoring mechanisms, and an additional 40% could be contributed to an inefficient layout. The Improve section targets these two areas.

A failure mode effects analysis was conducted. The Six Sigma team analyzed the data and system to determine the failure modes, potential effects, severity, occurrences, and detection. The key findings from the study are listed below and the actions taken are listed in the Improve section of this case study.

13.3.4 Improve

The fourth step of the DMAIC cycle focused on identifying and implementing system improvements based on the findings of the Analyze step. After identifying and analyzing the potential vital Xs, the team created a list of improvement initiatives to address each issue. Meetings and focus groups were held with the customers, managers, line workers, suppliers, and engineers and other staff to review the results and develop the following list of improvements. The following improvements were identified:

- Redesign the facility layout for optimal efficiency by reducing employee and material movement within the work area.
- Improve and monitor labor utilization versus the established engineering standards to ensure that savings are being realized.
- Standardize and document all the new processes.
- Train all employees and supervisors in the new processes.

13.3.4.1 New Facility Layout

After analyzing the current-state facility layout and the amount of traffic that overlaps, the team recognized that there was significant waste throughout. Through the analysis and application of facility planning tools, an improved facility layout greatly reduced movement in the facility.

With the information that was gathered a future-state layout was created. This was completed to minimize the amount of travel in the facility, decrease the cost of transportation for parts and products, cater to the proposed supermarket, and improve the overall tidiness of the facility layout. Figure 13.5 displays the improved future-state layout. By reducing the

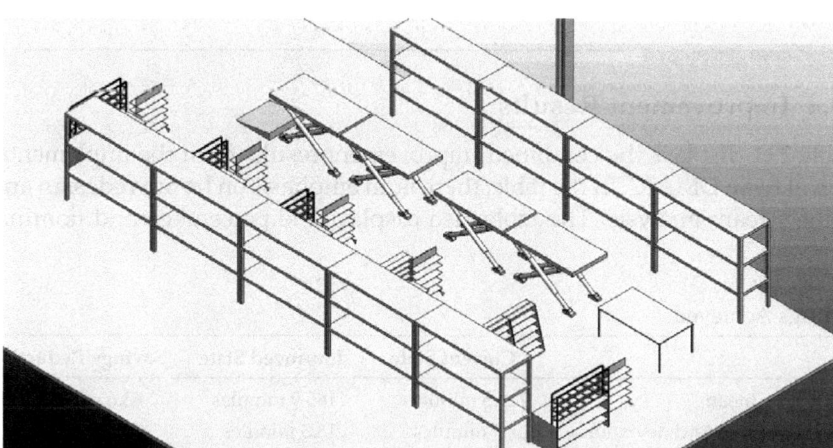

FIGURE 13.5
Improved future-state layout.

amount of space that the work unit occupies, the material and information can flow more efficiently.

13.3.4.2 Labor Utilization

Labor utilization based on engineering time standards was very low for the entire facility. Most operations were less than 60% efficient based on these standards. A key reason for this was lack of manager attention to these standards and outdated specifications. Several newer product lines also did not have labor standards in effect. Many work cells also had very low utilization rates and performed similar operations as other cells.

13.3.5 Control

The fifth and final step of the DMAIC cycle involves developing a system to monitor and sustain the improvement initiatives. This was primarily accomplished through the development of standard operating procedures, enhanced training, and daily progress tracking for all categories of employees within the work unit. The main purpose was to ensure the implemented improvements will hold and not revert to baseline figures. Logbooks were maintained within work centers to record key metrics. Graphical charts displaying the planned versus actual cost for each operation were also displayed daily as well. In addition, weekly meetings with the key staff were held to discuss performance and to generate new improvement solutions. These meetings involved a review of process times, customer surveys, and adherence to standard operating procedures.

13.4 Improvement Results

Table 13.1 displays the combined improvement results from the implementation of Lean DMAIC. In the table, there is an emphasis on layout redesign and value stream analysis. The table also displays the percentage and nominal

TABLE 13.1

Savings Achieved

	Current State	Improved State	Savings/Reduction
Cycle time mean	229.5 minutes	185.9 minutes	43.6 minutes
Cycle time standard deviation	27.0 minutes	13.3 minutes	13.7 minutes
Floor space requirements	1,120 square feet	810 square feet	310 square feet
Labor cost per unit	$160.10	$129.60	$30.40
Annual labor costs at 8,000 units	$1,280,000	$1,036,800	$243,200

savings achieved by rearranging the layout of the work unit to establish a more streamlined operation. The savings were accomplished by modifying the layout and work processes based on the spaghetti diagram, travel distances, and standard operation times. The savings were calculated by comparing the distances and travel times in the original layout and process to the revisions identified during the improvement phase.

By modifying the work unit layout the company can expect savings of over $243,000 per year based on cycle time/labor cost reduction at 8,000 units per year. This also equates to nearly a 20% increase in capacity.

13.5 Conclusions

Strong efforts are needed for organizations to remain viable in today's volatile financial climate. Lean Six Sigma with an emphasis on value stream analysis is such a tool to achieve this goal. This case study proposed a work unit layout redesign process based on the DMAIC cycle and demonstrated its application. It is suggested that organizations needing to trim costs and boost productivity could achieve similar result improvements for stakeholders and customers. Furthermore, the implications of this case study encourage other organizations to implement similar projects.

The Lean Six Sigma process and value stream analysis may be used as a guide or road map for Six Sigma development training and improvement. The case study demonstrated that through focused efforts, organizations can achieve great successes with relatively short project launch periods. A direction for future research would be to apply the value stream analysis methodology for different industries and gauge the positive and negative impacts.

Reference

1. George, M. (2002). *Lean Six Sigma: Combining Six Sigma Quality with Lean Speed.* New York: McGraw-Hill.

14

Case Study 3: High-Volume Printing Operations

14.1 Introduction

The company being studied was a small printing and visual arts company based in northwest Ohio, USA with 120 employees and annual sales of USD $1.5 million. An issue that this company was facing was the rising costs associated with acquiring the required capital to remain competitive in the printing business. They recently introduced all digital equipment into their facility.

In the last few years, the company entered into the business of printing sample boards for shingles, wood samples, brick samples, and similar products. These sample boards have a very high resolution and quality that could not have been achieved without the current scanning and printing technology. The company is exploring methods to create a market for these boards to reduce their customers' need for actual samples. Printing these boards is currently lower cost and more environmentally friendly than using actual samples for the products. As with any new product that has the potential to save money, the market is growing rapidly and the company currently does not have the capacity to handle the projected demand for the upcoming year.

In terms of production efficiency, the existing process for manufacturing the sample boards was not reaching its potential. The principles of Lean and Six Sigma are useful to the manufacturing process in order to maximize the production output while reducing the costs associated with defective parts [1]. Additional benefits include increased utilization of equipment and labor, improved safety, maximization of floor space utilization, and reduced inventory. These benefits will allow the company to increase its competitive advantage by allowing them to reach their projected production volume with minimal capital expenditure.

This case study follows the five-step methodology used in the Six Sigma process, the DMAIC cycle.

14.2 Application of Lean DMAIC Process

14.2.1 Define

The goal of this case study was to reduce process waste and increase production capacity to meet projected market demand for the new sample board product line. Six Sigma and Lean practices were applied to ensure that the changes would be effective and sustainable.

14.2.2 Measure

14.2.2.1 Delay Study

Currently there are two digital printing machines that are used to print the sample boards. These machines have the maximum capacity of about 20 boards per hour. Assuming three shifts per day and five days per week, the most boards the company could produce per year is about 75% of their projected yearly demand. These calculations do not allow for any unexpected delays or take into account the scrap rates. In addition, the company takes on other unrelated projects that utilize the printers on which the sample boards are printed. Various studies have been performed to determine the true capacity. These studies show that they will be unable to handle the forecasted demand.

In order to reach the maximum production capacity of the system, the major sources of waste need to be addressed. One of these issues is the quality of the incoming material used to make the sample boards. These blank white boards are shipped into the facilities where they are stored in inventory. These boards are already precut and are ready to be printed on when they arrive in the facility. Although the cost associated with inferior raw material is not a substantial issue (material with defects is simply not used), the disruptions caused by such quality issues are reducing the capacity of the process. The operator has to take time to inspect the boards to verify there are no scratches, discolorations, or warping. While he is performing this inspection and moving the defective parts to a scrap pile, the printing machine is standing idle.

Another concern in the current process is the inconsistent nature of the printing machines. Delay studies and feedback from the machine operator indicate a variety of reasons the printer would be offline. Some of these issues include a blown fuse on the computer that controls the printers, inconsistencies with the print color and quality between machines, downtime while the ink is being replaced, and general software issues that make it necessary to restart the computer. Although correcting the printer issues will greatly improve the overall productivity of the process, it was found that the only preventable source of significant downtime comes from replacing the ink.

A third area of improvement would come from the work design of the process. Currently, there are many times when the operator is waiting for the printers to finish. Also, there are times when the printer is sitting idle and the operator is still performing setups. One of the ways to adjust this is to design a specific set of instructions that optimizes machine utilization without sacrificing operator effectiveness.

14.2.2.2 Process Diagram

Two types of flow associated with the printing process of sample boards have been identified. One is the flow of information and the other is the material flow. The combination of both gives the general flow of the printing process. Although several areas of the information flow can be addressed for optimization the main focus will be the material flow.

The material flow describes the passage and transformation of raw material (blank sample boards) into finished product through the facility. The boards arrive stacked on a pallet from the warehouse and are stored in the production area, close to the printer. Raw material inspection takes place immediately before placing the board on the printer followed by surface cleaning. After printing takes place on the first surface, the boards are turned over, the other surface is cleaned, and then printed. The process is finalized with barcode scanning of the boards, final inspection, packaging, and depositing of the packed material in the shipping area. See Figure 14.1.

14.2.2.3 Time Study

Time studies were performed on both first and second shift operators, and the data obtained have been used to identify and simulate different situations. The data collected are shown in Tables 14.1 to 14.3 displaying the time in between steps 1 and 2, steps 2 and 3, and so on. Table 14.1 is data collected from the first shift operator, Table 14.2 is data collected from the second shift operator, and Table 14.3 represents data collected from operators 1 and 2 and averaged together. All data are presented in minutes.

The time between steps 1 and 2, 2 and 3, 4 and 5, and 5 and 6 basically consist of machine runtime only. For these steps, the operator has little effect on the speed of the process. However, the time in between the steps 3 and 4 and steps 6 and 1 involved the operator restarting the machine. Therefore there is more room for variation. Although little variation is seen between steps 3 and 4, it can be observed by looking at the standard deviations of the times between step 6 and step 1. The larger deviation in the time between cycles 6 and 1 is due to the machine cycle ending and the operator sometimes packaging the material instead of immediately restarting the machine for the next part, and sometimes immediately restarting the machine before packaging the boards.

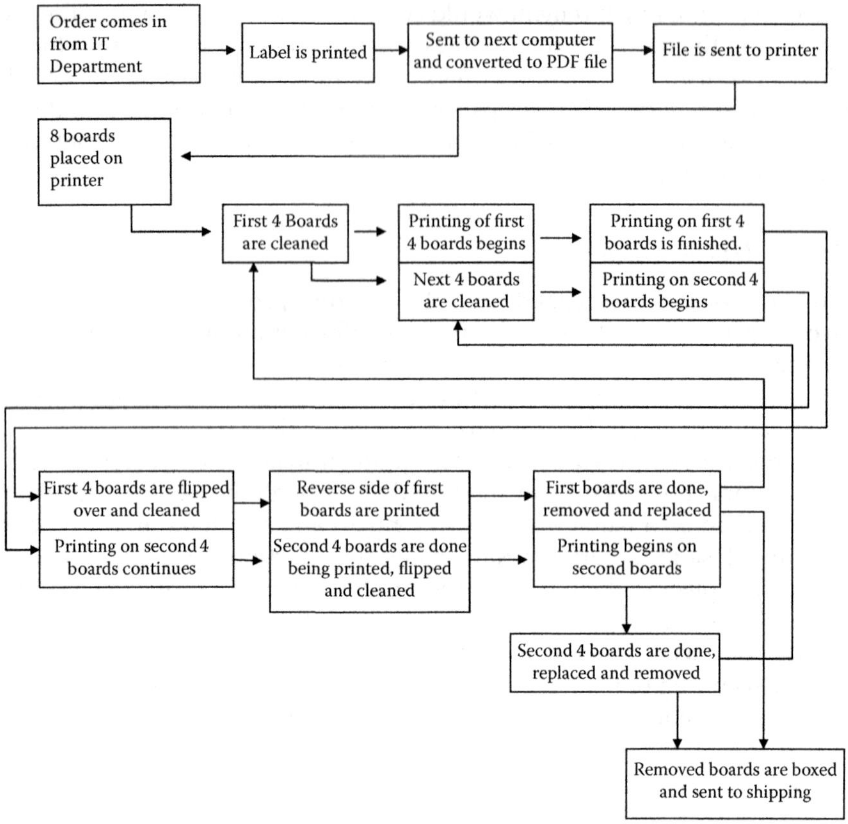

FIGURE 14.1
Sample-board process diagram.

TABLE 14.1
Process Time

		1 to 2	2 to 3	3 to 4	4 to 5	5 to 6	6 to 1	Process Time
Operator 1	AVG	4.95	5.66	0.93	5.09	5.59	2.29	24.18
	STDEV	0.17	0.08	0.28	0.05	0.07	1.66	2.07

TABLE 14.2
Process Time

		1 to 2	2 to 3	3 to 4	4 to 5	5 to 6	6 to 1	Process Time
Operator 2	AVG	5.29	5.36	1.39	5.68	5.27	3.37	26.08
	STDEV	0.37	0.48	0.45	0.81	0.46	2.99	3.43

TABLE 14.3
Process Time

		1 to 2	2 to 3	3 to 4	4 to 5	5 to 6	6 to 1	Process Time
Operators 1 and 2	AVG	5.12	5.51	1.16	5.38	5.43	2.83	25.13
	STDEV	0.27	0.28	0.36	0.43	0.26	2.32	2.75

14.2.2.4 Work Sampling

A work sampling study was performed on the operator to get a better understanding of the distribution of tasks that he performs. This was done by recording what action he was performing once a minute, on the minute. The duration of the study was just over two hours, giving 121 data points. Results and the analysis can be seen in Table 14.4 and Figure 14.2, respectively. These data were then analyzed and Pareto diagrams were created. One of the methods of analysis was to determine which activities were value-added and which ones were non-value-added as displayed in Figures 14.3 and 14.4. This was useful in finding improvement opportunities to reduce wasted time.

Analysis of the results showed that more than 30% of the operator's time was spent performing non-value-added tasks. The two major activities that

TABLE 14.4
Work Sampling Study Results

Activity	Count	Cumulative	Value Time	NV Time	Percent (%)
Working at the computer	30	30	30	0	24.79
Cleaning boards	23	53	23	0	19.01
Walking	16	69	0	16	13.22
Waiting	9	78	0	9	7.44
Set down blanks	8	86	8	0	6.61
Transporting blanks	6	92	0	6	4.96
Flipping boards over	5	97	5	0	4.13
Preparing next order	5	102	5	0	4.13
Scanning barcodes	4	106	4	0	3.31
Talking to coworkers	4	110	0	4	3.31
Packaging	3	113	3	0	2.48
Pick up completed parts	3	116	3	0	2.48
Quality check of blanks	2	118	2	0	1.65
Recording orders	1	119	1	0	0.83
Break	1	120	0	1	0.83
Reload spray bottle	1	121	0	1	0.83
Totals	**121**		**84**	**37**	
			69.42%	**30.58%**	

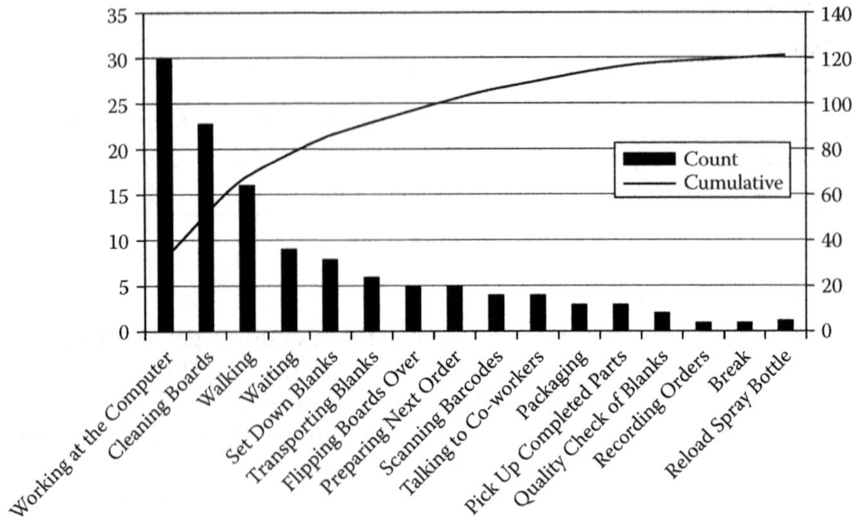

FIGURE 14.2
Work sampling study results.

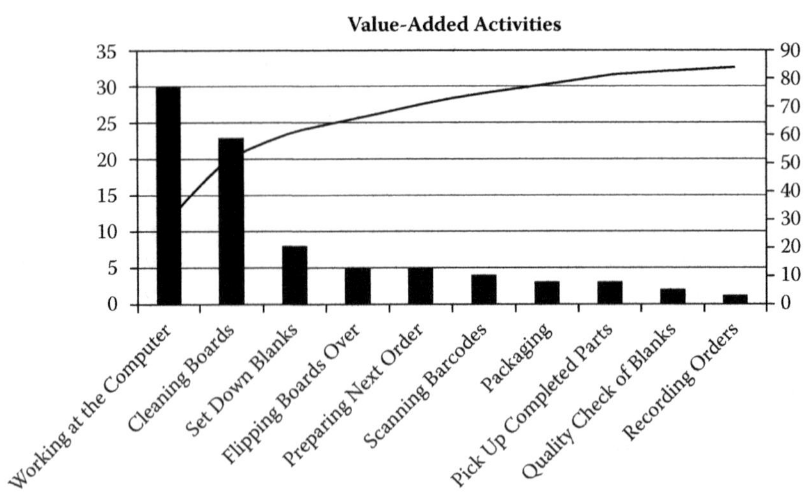

FIGURE 14.3
Value-added operator activities.

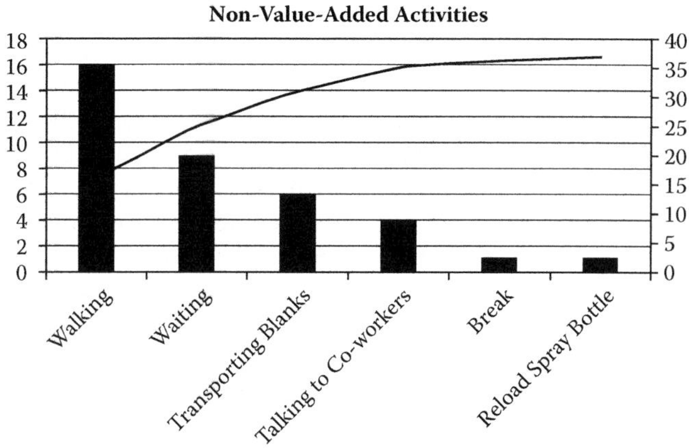

FIGURE 14.4
Non-value-added operator activities.

contributed to this were walking to or from the printer and waiting for the printer to finish a batch. Most of the time he spent walking was from the computer to the printer, but there was also a significant amount of time spent walking to pick up raw materials. In addition, it can be seen that the operator spent 19% of his time cleaning the blank sample boards. Improvements in this process would allow for more time to perform other tasks, such as running a second machine.

14.2.3 Analyze

14.2.3.1 Layout Evaluation and Material Storage and Material Handling

The main focus of this aspect of the study was the minimization of inventory and non-value-added work such as transportation and storage. For the redesign of the layout, the appropriate number of machines and operators resulting from the time study analyses were considered. Accurate measurements of the workspace were taken and a digital drawing of the current workspace was created. The measurements were then used to create floor plans that would allow three or four machines to operate in the current work space and minimize the amount of necessary operator movement between machines, raw material, and the packaging area.

14.2.3.2 Man–Machine Chart

The creation of a man–machine chart will be essential in the organization of the manufacturing process. A man–machine chart will highlight the tasks that need to be performed by the machine operator during the process, and the time that those processes will take. Those data will then be organized into a set of sequential tasks that outline the work necessary for the operator to perform in order to accomplish all the necessary jobs in the optimal way.

By outlining the work necessary for the operator to perform, the work and the order in which it should be performed can be standardized. By "scheduling" the work that the operator performs during the process, short breaks in the operator's process can be found. In order to prevent this time from simply being lost, work can be assigned to the operator to complete during his short breaks. In some cases, there are such large breaks in the process, the operator can run a second machine, or two operators can run three machines.

14.2.3.3 *Standardized Work*

From the data collected, a standardized work plan can be created to control how materials move through the process. Machining time, movement of raw material, facility layout, and operator actions are all factored into the creation of the standardized work plan. The machining time is an essential part of determining process time. Because of the importance it plays in the process, other factors have been created and modified around it. The facility layout has been designed to minimize the movement necessary for the operator. By reducing the movement of the operator between the machines, the raw materials, the computer, and the packaging area, the amount of time wasted in the transport of material will be minimized. Operator actions will also be taken into account in determining a standardized work plan because the operator plays a crucial role in how his or her workstation runs. The order in which the necessary tasks are performed in completing the process plays a crucial role in the efficiency of the process.

14.2.4 Improve

14.2.4.1 *Production Configuration*

One of the main purposes of the project is to provide guidance for capital investment. As demand is expected to rise in the future, the company will need to invest in additional equipment and labor in order to satisfy demand. With the current configuration, the company has an ideal output of 143,400 sample boards per year based on the following calculations:

$$(Hr \times Sl \times S \times D) \times R \times M = \text{Maximum Annual Output}$$

$$[15 \times 7.5 \times 3 \times 250] \times 0.85 \times 2 \cong 143{,}400\text{pcs/year}$$

where

Hr = maximum machine hourly production rate.

Sl = shift length, in hours.

S = number of shifts per day.
D = number of days worked in a year.
R = reliability.
M = number of machines used in the process.

These calculations represent an ideal maximum output based on equipment cycle time and reliability. In the near future the company expects order levels to exceed 200,000 sample boards per year and thus will need to invest in additional equipment and labor. Table 14.5 presents different production configurations and their maximum annual output. In making the recommendations, the main emphasis was placed on maximizing machine utilization and labor utilization. Labor costs for the recommended configurations were compared with the current costs in Table 14.6. When a cost baseline was established at the beginning of the project, the company was spending $2.105 per sample board. As shown in Table 14.6, the company could achieve the same output levels and reduce costs by 15% to $1.790 per sample board by applying Lean Six Sigma principles.

14.2.4.2 Plant Layout

The current production layout makes use of two printing machines (Figure 14.5). In the future the customer will be required to invest in additional equipment according to the increase in demand. As new equipment will be brought in, the opportunity to affect productivity positively through layout redesign will arise.

14.2.4.2.1 Three Machines Production Layout

If three machines are required to satisfy demand then the layout in Figure 14.6 is proposed. This layout was chosen based on the work configuration of three operators and no helpers considering the space available. The three-operator allocation results in the lowest price per unit for the three machines. Fortunately, the standardized work design did not change for the three operators. The operators were required to offset the machine initializations for printing to ensure that the computer was available when needed. This proposed layout requires no modifications of the production space, assures proper safety distances exist, and the flow of materials is kept to a minimum.

14.2.4.2.2 Four Machines Production Layout

If four machines are required to satisfy demand, the layout in Figure 14.7 is proposed. This layout requires some modifications to the production area. This layout also takes into consideration safety and efficiency. The optimal way to run this process with four machines is with two operators and two helpers.

TABLE 14.5

Labor and Machine Utilization for Different Production Configurations

Maximum Annual Output	Number of Machines	Labor Configuration	Task Distribution	Time Needed Per Cycle (s)	Labor Utilization[c] (%)	Machine Utilization[c] (%)	Machine Reliability[c] (%)
75,400	1	1 Operator	All tasks to be done by the operator	1,610	100	89.4	85
150,800	2	2 Operators	All tasks to be done by the operators	1,610	100	89.4	85
160,300		1 Operator – 1 Helper	Operator[a]: Helper: Remaining tasks	1,440 2,200	100 100	95	85
226,300	3	3 Operators	All tasks to be done by the operators	1,610	100	89.4	85
253,100		1 Operator – 3 Helpers	Operator[b]: Helper: Remaining tasks	900 1,310	62.5 90.9	100	85

301,700	4	4 Operators	All tasks to be done by the operators	1,610	100	89.4	85
337,500		1 Operator – 4 Helpers	Operator[b]: Helper: Remaining tasks	1,200 1,310	83.3 90.9	100	85
320,600		2 Operators – 2 Helpers	Operator[a]: Helper: Remaining tasks	1,440 2,200	100 100	95	85
377,100	5	5 Operators	All tasks to be done by the operators	1,610	100	89.4	85
421,800		1 Operator – 5 Helpers	Operator[b]: Helper: Remaining tasks	1,500 1,310	100 90.9	100	85

Note: Task times are expected to decrease due to work specialization and elimination of non-value-added activities.

[a] All inspections, raw material inspections, load data, start machine cycle, print labels.

[b] Load data, print labels (300 seconds/machine).

[c] Based on observed performances.

TABLE 14.6

Labor Cost for Different Production Configurations

Number of Machines	Labor Configuration	Comments	Annual Labor Cost ($)[a]	Labor Cost per Unit ($/unit)[b]
1	1 Operator		135,000	1.790
2	2 Operators		270,000	1.790
	1 Operator – 1 Helper	Best solution	243,000	1.516
3	3 Operators	Best solution	405,000	1.790
	2 Operators – 1 Helper	[c]		
	1 Operator – 2 Helpers	[c]		
	1 Operator – 3 Helpers		459,000	1.814
4	4 Operators		540,000	1.790
	3 Operators – 1 Helper	[c]		
	2 Operators – 2 Helpers	Best solution	486,000	1.440
	1 Operator – 3 Helpers	[c]		
	1 Operator – 4 Helpers		567,000	1.769
5	5 Operators		675,000	1.790
	4 Operators – 1 Helper	[c]		
	3 Operators – 2 Helpers	[c]		
	2 Operators – 3 Helpers	[c]		
	1 Operator – 4 Helpers	[c]		
	1 Operator – 5 Helpers	Best solution	675,000	1.600

[a] Based on an annual salary of $45,000 for an operator and $36,000 for a helper; three shifts per day.

[b] Found by dividing the annual labor cost by the maximum annual number of units (from Table 14.5).

[c] Tasks allocation and work equity makes this configuration not feasible.

14.2.4.3 Man–Machine Chart

From the data, it was found that the average operator only works for about 13 to 14 minutes out of the 21 to 22 minutes that it takes for the printer to complete the process. This means that if the operator were operating two machines at once, he would be busy for 21 to 22 minutes of the machine operation and his downtime would be eliminated.

Assuming that there is sufficient demand for the sample boards, it may be necessary to run more than two machines. The company's current capacity was calculated based on the assumption that two machines run full time. The company should fall short of producing their projected forecast of 200,000 boards annually. Additionally, this is assuming that both machines can be dedicated to producing only sample boards, which is not the case. If the company's demand does meet its projected forecast, it will have to invest in a third machine in order to dedicate production on two machines entirely to sample boards.

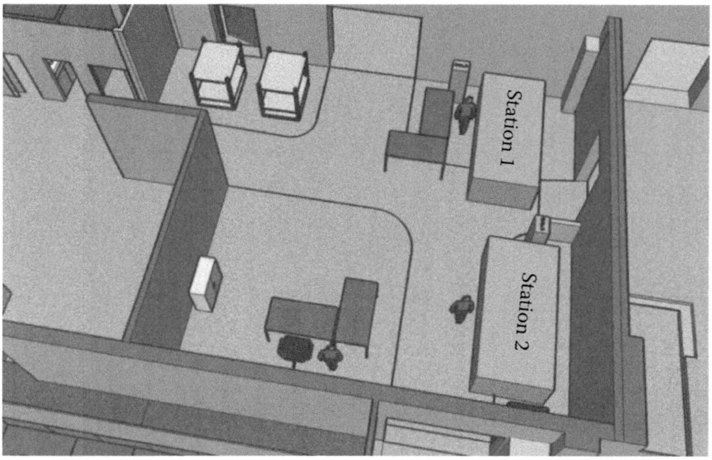

FIGURE 14.5
Current layout configuration.

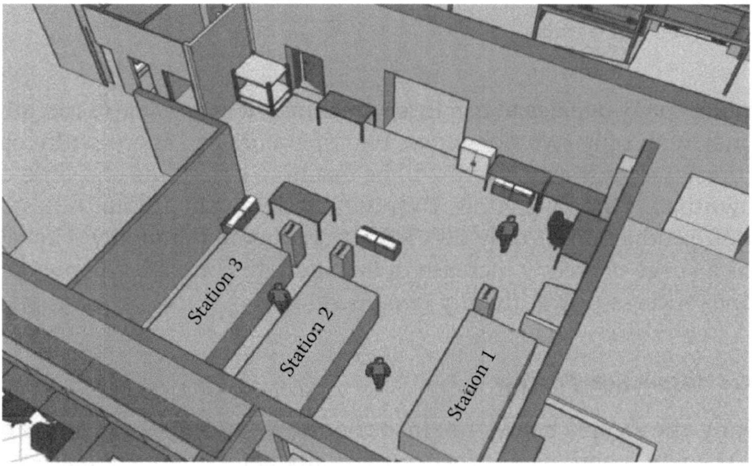

FIGURE 14.6
Proposed layout: three machines.

While running two machines, the management will have to decide whether time is their most important constraint or if money will be their critical constraint. If they decide to conserve money, and use one operator, they will be sacrificing machine time because the machines will be forced to sit idle while the operator performs other functions. If they decide to operate both machines with two operators, the operators will sit idle while the machine prints. So, the number of operators on each machine will be up to the discretion of the management at the company.

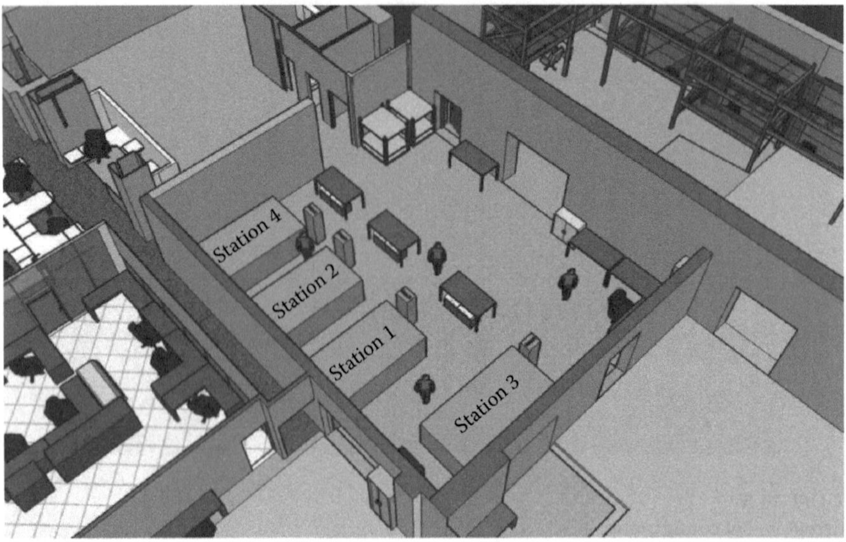

FIGURE 14.7
Proposed layout: four machines.

If the company decides to run three machines, it is possible to run all three machines with only two operators. Two operators can successfully operate three machines, however, there will be machine downtime and little operator downtime. Three machines running continuously are not necessary to maintain production of only 200,000 sample boards annually. However, if there is a surge or steady increase in demand, it is possible to operate three machines successfully with only two operators.

14.2.4.4 Inspection Process

Currently, the sample boards are inspected twice during the printing operations. The raw materials are checked visually before they are placed on the printing machine. These white boards are inspected not only for surface finish flaws, such as scratches, chips, and flakes, but also for the coloring of the board. If there are any color imperfections, the printed color of the sample image will not show up accurately, as the computer is balanced to print on pure white boards. Another imperfection that the operator is examining the board for is any warping or bending of the sample board. If the board does not sit flat on the printing table, the suction used to hold the board in place will not hold it effectively. If the boards have edges that are bent up, there could be inadequate clearance between the boards and the printer head, causing discoloration, smudging, and potential damage to the printer.

The other inspection process comes at the end of the operation. Because the company is competing mainly on innovation and quality of material, the sample boards must fully mimic the appearance of the sample. Any problems with the printer head will cause ink blotches, so routine maintenance needs to be performed on the printer. In addition, at the beginning of every shift, the operator will run a quality check to make sure that the colors are balanced correctly.

Unfortunately, because quality is such an important aspect of this process and it is extremely difficult and expensive to create a gauge that can measure the surface quality of this type of product, the human eye is really the only viable option for maintaining the quality of the product in a cost-effective manner. Addressing the issues associated with the warping of the raw material (i.e., causes, suppliers, storage, etc.) is beyond the scope of this project and unable to be addressed at this time.

14.2.4.5 Implementation of Kanbans

An important determinant of the success of "push" production scheduling is the quality of the demand forecast which provides the "push." Kanban, by contrast, is part of a pull system that determines the supply, or production, according to the actual demand of the customers. In contexts where supply time is lengthy and demand is difficult to forecast, the best one can do is to respond quickly to observed demand. This is exactly what a kanban system can help: it is used as a demand signal that immediately propagates through the supply chain. This can be used to ensure that intermediate stocks held in the supply chain are better managed and are usually smaller. Depending on the number of machines used in the process, different reorder points will exist for printing supplies.

E-kanban will make use of self-adhesive labels printed with a barcode representing the stocked item. Labels will be attached to the received supplies before storage in the production area. After the materials have been taken from the storage area the kanban label will be removed by the operator and handed to the stockkeeper who will scan it into the computer database. E-kanban has the advantages of being simpler and more reliable; the stockkeeper will not have to enter the values manually but can scan them with a barcode scanner. The short amount of time necessary for this operation will also ensure promptness in which stocks are handled. Table 14.7 gives levels at which the stock should be held to optimize the E-kanban system.

14.2.5 Control

In order to ensure that the proposed methods of improvement are sustained, the company will need to implement a set of control systems. These control

TABLE 14.7
Printing Supplies Reorder Point

Number of Machines	2		3		4	
	Point	Quantity	Point	Quantity	Point	Quantity
Alcohol (gallons)	1	2	1	2	1	2
Ink: C, M, Y (each)	2	2	2	2	2	2
Ink: Black	2	2	2	2	2	2
Rags (lb)	3	10	4	10	5	10
UV light bulbs	2	2	2	2	2	2
Label stickers (boxes)	1	3	1	3	1	3

methods will also aid the company in being more responsive to process variations in the future and will be better equipped to handle unexpected deviations. There are several tools that would be of use to the company given their circumstances.

One useful tool would be a standard operating procedure (SOP) for the printer operators. As shown in this study, the operators do not necessarily use the same methods to print the sample boards, which creates costly delays in the printing process. A SOP consisting of very clear operations will be useful for reducing or eliminating the variations in operator procedure.

Another tool that is useful for controlling the system is a check sheet. Every time there is an unplanned delay in the process, the operator should mark the delay with the time and date. This can be useful in planning for the delays. For example, the operations manager at the facility might notice a trend that the ink jet is more likely to misfire toward the end of the life of the ink cartridge. This would be useful in anticipating printer misfires.

The overall goal of this project was to make the sample board printing process more efficient in order to increase the company's competitive advantage. It has been shown that the proper application and implementation of Lean and Six Sigma techniques can be used to create a better process that is more cost effective and can meet the demands of the customers [2]. After defining the problem, it was clear which performance measures needed to be studied. From these data, improvement opportunities were analyzed and action plans were created to help implement those plans. Finally, a standardized method was developed in order to ensure that the process would be sustainable. If the company is able to carry out the proposed suggestions successfully, they will be able to determine the optimal number of employees to minimize the labor costs per unit produced and achieve labor cost savings of 15%. In addition, the proposed improvement methods will decrease the number of production defects and improve the overall quality of the finished goods through implementation of standard operating procedures and the E-kanban system.

References

1. Henderson, K.M. and Evans, J.R. (2000). Successful implementation of Six Sigma: benchmarking: General Electric Company. *Benchmarking: An International Journal*, 7(4): 260–281.
2. Park, S.H. (2002). Six Sigma for productivity improvement: Korean business corporations. *Productivity Journal*, 43(2): 173–183.

15

Case Study 4: Material Recovery and Recycling Processing Facilities

15.1 Introduction

The purpose of this case study was to understand better the application of Lean Six Sigma principles in government-controlled operations for municipal recycling collection and processing. Specifically, the objective of the project was to improve the processes and operations for a material recovery facility (MRF) located in Toledo, Ohio, USA that was operated by the local government. This included aligning and optimizing processes and removal of process-generated defects and errors. These improvements primarily focused on the reduction of municipal solid waste (MSW) entering local landfills, optimization cycle times for the MRF, enhancement end user/customer satisfaction, efficiency improvements, cost reduction, and the elimination of errors/defects. The intent of these objectives were twofold: first to demonstrate that Lean Six Sigma techniques can be successful applied in the material recovery industry and second, that these techniques can be successfully applied in a government controlled operation.

Subobjectives included developing process standards to eliminate bottlenecks and redundancies. Lean Six Sigma was selected for this study to address both cost and quality issues that the facility was experiencing. Due to lower tax revenues, the government-operated facility was faced with cost reductions to meet budget targets at the time of the study. In addition, recycling mills and processors that received the end product from the facility were dissatisfied with the level of contamination in paper and plastic bales. This project was chosen to address these issues by lowering operating costs and enhancing customer satisfaction by reducing defects. At the time of this study, the MRF administration team was preparing to add additional equipment to process additional recyclable materials collected from nearby cities and towns. The MRF administration team established a goal to become more productive and efficient to utilize tax dollars and serve the public better.

15.2 Application of Lean DMAIC Process

15.2.1 Define

The goal of this case study was to reduce process waste, reduce defects, and increase production capacity to meet projected increased capacity needs for the Lucas County MRF. Six Sigma and Lean practices were applied to ensure that the changes will be effective and sustainable. To begin the study, a project charter was created that identified the project team, goals stated above, start date, and expected end date.

15.2.1.1 About the Organization

A MRF is an operation where recyclable materials are received, sorted, and prepared for sale to end users or buyers to create new raw materials. The MRF operation analyzed for this study involved a facility that employs 30 and is owned and operated by the local government. The Lean Six Sigma team consisted of the facility manager, supervisor, baler operator, and a maintenance employee. This project represented the first initiative related to Lean Six Sigma for the facility.

As displayed in Figure 15.1, a clean MRF process commingles recyclables generated by local residents and businesses. MRFs are designed such that the input materials or recyclables from curbside recycling programs can be separated and processed into marketable commodities. The MRF discussed for this case study is a basic multistream MRF with manual sorting operations.

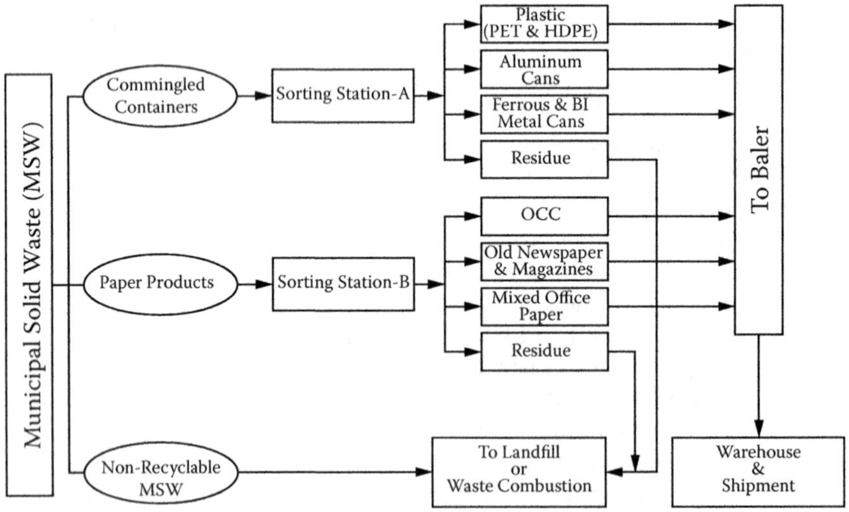

FIGURE 15.1
Flowchart for a basic clean MRF operation.

As displayed in Figure 15.1, a multistream MRF separates and bales commingled containers including ferrous metals, aluminum cans, polyethylene terephthalate (PET) and high-density polyethylene (HDPE) bottles, mixed glass, old corrugated containers (OCC), mixed office paper, old newspapers, and magazines. Nonrecyclable and residue from both streams after manual sorting are collected and sent to landfill. After the manual sorting at Sort stations A and B as shown in Figure 15.1, these materials are baled using a baler and are finally ready for shipment to the end users.

This case study examines a MRF located in Toledo, Ohio, USA. The MRF is owned by a division of the local government, the Lucas County Solid Waste Management District (LCSWMD). At the time of the study, the MRF was used to sort and sell approximately 10,000 tons of recyclable material annually. The LCSWMD funds and maintains approximately 174 drop-off recycling sites throughout Lucas County. The drop-off sites are located at apartment complexes, schools, large grocery stores, and metro parks. Each drop-off container is owned and maintained by the LCSWMD. Daily trucks are dedicated to pick commingled fiber and container streams. As the trucks reach capacity they unload the material at the MRF for sorting and baling. Before unloading materials trucks are weighed and recorded for payment and in order to keep track of the system. A ticket is generated for each truck and total weight is recorded. The facility has a sorting center, a baler, and a warehouse for shipping final products to end users. Figure 15.2 displays a picture of the MRF sorting process.

FIGURE 15.2
Sorting operation for the Lucas County MRF.

In terms of process efficiency, the MRF operations were not reaching their full potential. Large variations in daily processing rates were observed in conjunction with large inventories of unprocessed materials. The principles of Lean and Six Sigma are useful to the manufacturing process in order to maximize the production output while reducing the costs associated with defective parts [1]. Additional benefits include increased utilization of equipment and labor, improved safety, maximization of floor space utilization, and reduced inventory. These benefits will allow the MRF to increase efficiencies by allowing them to reach their projected production volume to utilize tax dollars best. In addition, efficiency enhancements and variability reduction will allow the MRF to increase its capacity with minimal additional capital expenditure. This was a critical need for the MRF, inasmuch as plans were being developed to begin the processing of an additional 5,000 annual tons of recyclable materials from a local community's curbside collection program. The following sections discuss the baseline measures, including the DPMO, analysis, improvements, and control strategies.

15.2.2 Measure

The purpose of the measurement phase was to baseline the existing processes to establish current-state metrics [2]. Several weeks of on-site data collection at the MRF were required. Some of the tools used in the measure phase of the project were Pareto analyses, time studies, surveys, interviews, and check sheets.

15.2.2.1 Pareto Analysis

A Pareto chart is a vertical bar graph that displays problems or improvement opportunities in a prioritized order, so it can be determined which problems or opportunities should be addressed first. It is a tool used on the 80:20 rule which means usually 80% of the problems come from 20% of the processes. The categories are sorted in decreasing order from left to right in a Pareto chart by their count or cost, whichever is displayed [3]. Figure 15.3 displays the Pareto analysis for the amounts (in tons) of materials sorted per year at the MRF.

The annual recyclable material to be processed through the MRF was 11,432 tons. From the Pareto chart, the most collected material at the MRF was old newspaper (ONP), mixed office paper (MOP), and then old corrugated containers (OCC) and aluminum cans.

15.2.2.2 Time Studies

Time studies are the continuous surveillance of a job or process, using a timekeeping device (a stopwatch was used in this project) to record the time taken for accomplishing a job. After recording the time, the worker's

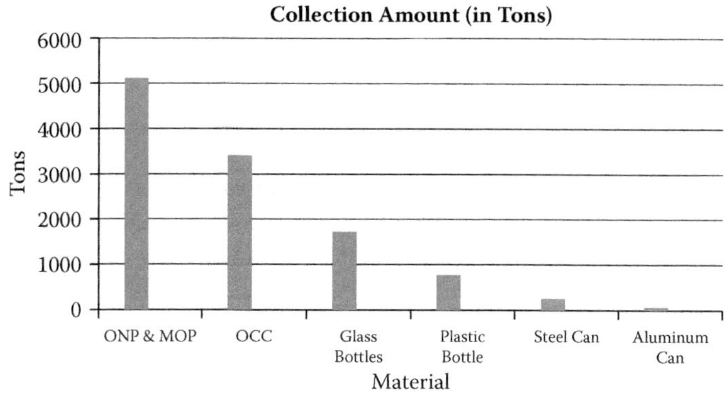

FIGURE 15.3
Pareto analysis.

TABLE 15.1

Cycle Time to Bale Mixed Office Paper (MOP) through the Baler

Time In	Time Out	Total Time For Process (Mins)
0.0000	0.0812	4.9
0.0812	0.2787	11.9
0.2787	0.3320	3.2
0.3320	0.3799	2.9
0.5400	0.6600	7.2
0.6070	0.6689	3.7
0.6835	0.7924	6.5
0.3799	0.5400	9.6
Average Std. Time		6.2

performance time (level) is recorded and then the data are used to make the standard time for the task or process. Standard times (minutes/cycle) for the baler were calculated from data observations for each material. A sample of the data collected for the cycle time to bale a load of MOP is displayed in Table 15.1.

Table 15.1 displays the average standard time taken for a bale of paper to process through the baler. The average time to process and bale the MOP was 6.2 minutes with a standard deviation of 0.95. The average time to process a bale of OCC was 6.7 minutes with a standard deviation of 1.10. The variation in the system was due to breakdowns of the cable system that covered the bale, removal of unwanted glass particles coming out from the baler, and other continuous maintenance operations. Time studies were also conducted to determine transportation cycle times to transport the bales from the baler to the warehouse staging area. The average cycle time to transport the bales was observed to be 1.8 minutes.

15.2.3 Analyze

To begin the analysis phase, a fishbone diagram was created to understand sources of variation and delay in the processes better as displayed in Figure 15.4. As shown in the figure, material, machinery, methods, and staff were the main factors for variation and delay. From a methods standpoint, employee surveys and interviews were used to collect data. These surveys revealed that the manual sorting method used could be improved to reduce the overall defects. Other comments included issues regarding complexity of the process, limited quality control, and lack of standard operating procedures.

The primary issue with the staff was related to unscheduled work absences: if a scheduled operator was not present it became very difficult for another operator to cover the shift, due to lack of standards and contingency plans. The staff did appear to require additional training regarding the defects in the process.

As displayed in Figure 15.4, raw material caused significant variability in the system. The main reason was related to broken glass as it could contaminate the end products. OCC is also one of the major defects in the raw material stream based on condition and quality. Shredded paper is another main defect because it is difficult to bale.

An additional source of variation in the process was related to machine breakdown and improper resource utilization. For example, a single baler

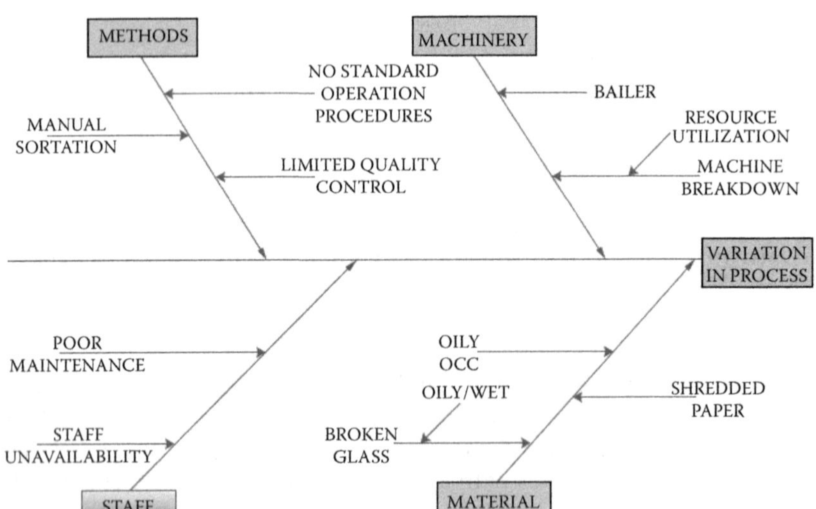

FIGURE 15.4
Fishbone diagram.

was used for all material types. Due to improper maintenance and extensive use of the baler, the cable system that wraps the bale tended to break and malfunction frequently. There was a large amount of non-value-added time lost in fixing the cable wiring system.

15.2.3.1 Machine Utilization Analysis

To understand the equipment usage better, the equipment requirements were determined via three on-site data collection audits. The amount of equipment required for a process is referred to as the equipment fraction. The equipment fraction can be determined for a process by dividing the total time required to perform the process by the time available to complete the process. The total time required to complete the process is a product of the standard time for the process and the number of times the operation/process is to be performed [4]. Standard time was then calculated by using time studies and other statistical tools.

$$F = S * Q * \left(\frac{1}{H * E * R} \right)$$

where

F = number of machines required per shift to perform the task.
S = standard time (minutes) per unit produced.
Q = number of units to be produced per shift.
H = amount of time available per machine.
E = actual performance, expressed as a percentage of standard time.
R = reliability of machine, expressed as percent "up time".

Additionally, equipment requirements are a function of the factors such as number of shifts per day, setup time, degree of flexibility, layout type, and maintenance. Standard cycle times were measured for current baler per material to determine resource utilization. As shown in Table 15.2, the baler was used to process 12,000 bales per year. The baler was estimated to be 85% reliable based on the machine utilization audit. The efficiency in terms of standard cycle time was measured to be 90% which means that machine was processing at 10% less than its standard. Based on standard cycle times and 250 working days of 8-hour shifts a day, the baler was utilized 85%.

15.2.3.2 DPMO

Defects per million opportunities (DPMO) are a measure of process performance or nonconformities per million opportunities. A defect is also known as a nonconformance of quality characteristic to its requirement and specification. For this study, defects were defined as tons of material by categories

TABLE 15.2

Equipment Requirement by Material for the Baler

Material	Average Cycle Time (Min)	Standard Deviation of Cycle Time	Annual Tons to Be Processed	Pounds per Bale	Bales per Year (Q)	Available Time per Year (H)	Machine Reliability (R)	Efficiency (E)	Machine Fraction (F)
ONP	6.2	0.95	3,759	1,630	4,612	120,000	85.00%	90.00%	0.31
OCC	6.7	1.10	3,410	1,700	4,012	120,000	85.00%	90.00%	0.29
MOP	6.2	0.80	1,356	1,630	1,664	120,000	85.00%	90.00%	0.11
PLASTIC	7.1	1.10	793	1,200	1,322	120,000	85.00%	90.00%	0.1
STEEL CAN	7.1	0.99	275	1,830	301	120,000	85.00%	90.00%	0.02
ALUMINUM CAN	7.1	0.95	82	1,830	90	120,000	85.00%	90.00%	0.01
TOTAL			9,675	9,820	12,000		0.85		

that were received at the MRF that could not be recycled due to improper condition or contamination. Defects for the processes are shown in Table 15.3 with the annual weights of defects. As displayed in the table, the main defects of the process are glass particles which amounted to 85 tons per year. Based on this analysis, DPMO for the process was calculated as 2,788 defects per million opportunities, which corresponds to a Sigma level of 4.3 in terms of weight.

TABLE 15.3

Annual Weight of Solid Waste or Defects from the MRF

Component	Tons per Year	Percent of Total (%)	Tons Recycled	Tons Not Recycled
Glass	31.85	94.1	0.00	31.85
Mixed office paper	1.37	4.1	0.00	1.37
Nonrecyclable/food	0.22	0.7	0.00	0.22
Cardboard	0.15	0.4	0.00	0.15
HDPE (2)	0.13	0.4	0.00	0.13
Aluminum cans	0.03	0.1	0.00	0.03
Newspaper	0.00	0.0	0.00	0.00
PET (1)	0.10	0.3	0.00	0.10
Magazines	0.00	0.0	0.00	0.00
Plastic wrap	0.00	0.0	0.00	0.00
Other	0.00	0.0	0.00	0.00
Other plastics	0.00	0.0	0.00	0.00
Total	**33.85**	**100.0**	**0.00**	**33.85**

(1) Plastic type 1
(2) Plastic type 2

TABLE 15.4

Total Revenue Generated from Recycling

Material	Dollar per Baled Ton	Annual Tons to be Processed	Annual Revenue
ONP	85	3,759	319,515
OCC	110	3,410	375,100
MOP	82	1,356	111,192
Plastic	180	793	142,740
Steel can	180	275	49,500
Aluminum can	180	82	14,760
Total		**9,675**	**1,012,807**

15.2.3.3 Cost Analysis

The annual utility/maintenance expenses for the MRF were calculated and estimated to be $95,000 and annual labor expenses of $820,000. A revenue analysis and an average dollar amount per ton for each recyclable material were also calculated and are displayed in Table 15.4. The total annual revenue generated from the sale of recyclables based on the average commodity rates was $1,012,807. This cost analysis served as an initial baseline for process improvements based on Lean Six Sigma concepts.

15.2.4 Improve

The next phase of the study involved improving the system to reduce variability and cost by removing major bottlenecks and sources of defects [5]. Value-added and non-value-added analyses were a major part of this phase. Table 15.5 displays the dollar loss per year due to residue in the system that

TABLE 15.5

Dollar Loss per Year due to Defects

Material	Tons per Year	Dollar per Baled Ton	Annual Dollars Lost ($)
Glass	31.85	25	796.25
Mixed office paper	1.37	82	112.34
Cardboard	0.15	110	16.5
HDPE (2)	0.13	180	23.4
Aluminum cans	0.03	180	5.4
Newspaper	0.00	85	0
PET (1)	0.1	180	18
Totals		**33.63**	**971.89**

(1) Plastic type 1
(2) Plastic type 2

cannot be sold on the commodity market, which amounted to nearly 35 tons of material and $1,000 per year. Upon reviewing the data, major improvement areas were broken glass particles, contaminated OCC, baler machine failures, time wasted in transportation of baled material to the warehouse, and loose shredded paper.

15.2.4.1 Value-Added versus Non-Value-Added Analysis

Lean manufacturing is a systematic approach to identifying and eliminating waste which is also known as non-value-added activities through proper continuous improvement techniques [6]. The three factors that were evaluated for work to be considered as value-added were [7]:

- *Capacity:* Machinery, resources, tools, and employees used in the process must have the necessary capacity to produce a final product adding value; for this study, utilization for the baler was considered.
- *Information/instructions:* Employees must know their final product and process to achieve the final product with minimum waste or non-value-added activities. For the MRF, all workers should have a thorough understanding of the overall process plus they should understand the difference between recyclables and nonrecyclables very well.
- *Materials:* The material given to workers should be free from defects and able to be processed and sold on the commodity market. A worker should know which raw material is acceptable or unacceptable.

The five sources of waste that were identified and analyzed for this study were adapted from previous literature [8] and included:

1. *Defects:* Defects are due to various reasons starting from material, inadequate training, labor, sorting, or deficient maintenance. For the MRF operation, defects were mainly due to baler breakdowns, defects in raw materials, and incomplete engineering specifications. At the MRF, the time studies indicated that 29% of the non-value-added time is due to repairs for the cable wire system which wraps the final baled products.
2. *Transportation:* Unwanted movement of materials within the facility makes it a waste or non-value-added activity from the customer's viewpoint. Several examples observed at the facility included multiple movements of MOP or OCC to several staging areas near the baler. A single movement from the unloading zone to the baler would reduce this form of waste.
3. *Waiting:* The time wasted such as waiting for raw material, waiting time for equipment or tool setups and instructions, waiting for

tooling, and waiting for workers all adds up to a large amount of non-value-added activities to the final product. The waiting total time observed during the time studies was 1.1 minutes per 8.1-minute cycle (13.6% for the total cycle time). Wait time issues included delays related to waiting for the next process to start, idle time between two processes, and time wasted during miscellaneous activities including restroom breaks. Wait time represented 47% of the non-valued–added activities per cycle.

4. *People:* Time wasted in not utilizing the abilities of a worker or employee to his or her fullest potential describes this form of waste. Morale and company culture is also one of the main factors for this kind of waste. One example of this form of waste observed at the facility included work activities of the trained baler operator. Several instances were observed that involved him doing lower-level work such as sweeping the floors or operating a fork lift.
5. *Motion:* Extraneous motion derives from poor plant layout and workplace design contributing to motion losses. It includes time wasted in searching for tools, extra product handling, or material stacking. For the MRF, time taken by a worker to move baled items from the baler to the shipping area was included in this type of waste.

The main waste or non-value-added activities in our case study were the times required to fix the baler cable system, which occurred seven to eight times a day.

15.2.4.2 Capacity

Due to changing demands of a product an organization needs to determine the production capacity. For the MRF, the decision to process additional recyclable material from nearby communities related directly to capacity. Capacity can be increased by using modern manufacturing methods, continuous improvement in the facility, optimum utilization of machines and workers, and many other process improvement techniques depending upon the business to increase efficiency and reduce waste. The formula used for calculating the tons per hour of a material processed at the MRF was:

$$T_x = B_x * W_x * \frac{1}{2000}$$

where

T_x = tons per hour processed for material x.

B_x = number of bales processed in one hour for material x.

$$B_x = N_x * 60 * \frac{1}{P_x}$$

TABLE 15.6

Tons per Hour Processed for Each Material

Number of Bales (N_x)	Material (x)	Total Time for Processing the Material (P_x) – (*MINS*)	Pounds per Bale (W_x)	Tons per Hour Processed (T_x)
9	Paper	58.6	**1,630**	**7.5**
5	Comingled	35.2	**1,515**	**6.5**
9	OCC	61	**1,700**	**7.6**

TABLE 15.7

Tons Processed per Hour After Recommendations

Number of Bales (N_x)	Material (x)	Total Time for Processing the Material (P_x) – (*MINS*)	Pounds per Bale (W_x)	Tons per Hour Processed (T_x)
9	Paper	54.6	1,630	8.1
5	Comingled	31.2	1,515	7.3
9	OCC	60	1,700	7.7

N_x = number of bales processed for material x in total time.
P_x = total time taken to process the material.
W_x = pounds per bale for material x.

Table 15.6 displays the capacity of the system under its current state in terms of processing capability per hour per material type. Tons per hour processed for MOP, containers, and OCC were 7.5, 6.5, and 7.6 tons per hour, respectively. The capacity of processing each material per hour can be increased by reducing the non-value-added time and waste from the process. By implementing the recommendations that were identified, the tons per hour processed at the MRF could reach a higher level. Table 15.7 represents the tons per hour that could be processed assuming the recommendations were implemented. After implementing the recommendations there will be at least an increase in tons processed per hour by 7.3% in paper bales, 12.8% in commingled bales, and 1.6% in OCC bales.

15.2.5 Control

The purpose of the control phase is to monitor and maintain improvements. Control charts, documentation, and communication are the critical components of this phase. To verify that improvements were achieved, a statistical analysis between the previous and improved system states was analyzed. To be statistically confident that the mean cycle time after the recommendations were implemented was lower than the baseline means cycle times, a hypothesis test was performed. A *t*-test was conducted with the following hypothesis for the mean cycle time of plastic bales:

Null hypothesis, $H_0 : \mu \leq 6.5$, new mean cycle time is less than or equal to 6.5.

Alternate hypothesis, $H_1 : \mu > 6.5$, new mean cycle time is greater than 6.5.

The sample size was 20 cycles for the each scenario (before and after recommendations). The normality assumption for the *t*-test was validated using a Kolmogorov–Smirnov test for each sample set. The *F*-test was used to validate the equal variance assumption between the two populations. The samples were assumed independent as they were taken before and after process changes. The test statistic for this test was calculated to be 0.559. At a 5% significance level, the critical *t*-value ($t\alpha$) is 2.776. The *t*-test indicates that we do not reject the null hypothesis. In other words it indicates there is a 95% confidence level that the new mean is less than 6.5 minutes after implementing the recommendations. Statistically, it can be concluded that new mean cycle times are lower than the previous-state cycle times.

15.2.5.1 Control Charts

To maintain and monitor the improved cycle times, control charts were created to identify special causes of variation, delays, or bottlenecks in the system [9]. The control charts were created using a standard Three Sigma control limit. A sample control chart for the cycle time to bale a load of OCC is displayed in Figure 15.5. The control charts utilized to monitor performance were the $\bar{x}$ and *R* charts. The $\bar{x}$ and *R* charts were chosen because a

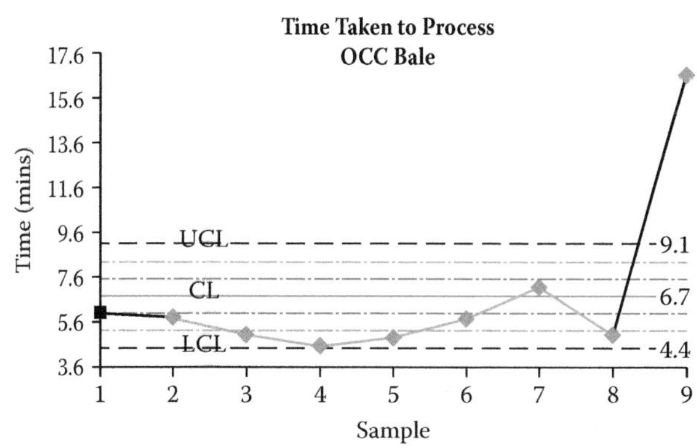

FIGURE 15.5
Control chart showing time taken to process the OCC bale.

process variable (cycle time) was being monitored for small and constant samples of five. The samples were collected every hour.

The control charts were automatically generated per cycle once the operator pressed the start button to begin a bale cycle. The control chart offered a real-time measurement for the leadership team at the MRF to monitor production rates and identify delays quickly. For example, in Figure 15.5 the final sample is out of control due to a baler wire jam. Special causes of variation were addressed by a team consisting of the operator, maintenance team member, and the shift supervisor.

15.2.5.2 Process Capability

The capability of the improved process was calculated by measuring the tons processed per hour for each material. The final capability is displayed in Table 15.8 and it indicated the before and improved levels. Costs related to poor quality were also significant. In terms of lost revenue, the costs of poor quality were approximately $1,000 per year as displayed previously in Table 15.5. The preventive costs to remove these defects (broken glass mixed in MOP of OCC) were minimal and addressed by proper handling and separation of glass materials.

15.2.5.3 Managerial Implications

The managerial implications from this study strongly indicate that lean process redesign had a significant impact on reducing operational costs (approximately $65,000 per year). Maintaining the improved standard cycle to process bales of material is critical to maintain these savings and the developed controls will allow the managerial team to accomplish this task. Also, the cost of poor quality related to broken glass could generate an additional $1,000 per year if proper separation and handling are enforced.

TABLE 15.8

Final Capability Chart

	Before	Improved Levels
Tons processed per hour for paper	7.5	8.1
Tons processed per hour for OCC	7.6	7.7
Tons processed per hour for commingled	6.5	7.3
Total savings (annual)	—	$65,019

15.3 Conclusions

This case study demonstrated that Lean Six Sigma can be successfully applied in the recycling and material recovery fields. In addition, it also demonstrated that these concepts can be applied with success in government-controlled operations. The case study illustrated how the DMAIC approach of Six Sigma aided in identifying bottlenecks and barriers to the MRF operations. DPMO and Sigma level calculations for the current process were calculated to identify potential economic and environmental benefits. By reducing non-value-added activities, productivity was increased by 7.3% for paper bales, 12.8% for commingled bales, and 1.6% for OCC bales. The case study also enhanced employee participation in process improvement, the use of statistical tools to solve problems, increased process knowledge, and it motivated the MRF leadership team toward Six Sigma approach.

Key lessons learned from this study relate to identifying, implementing, and maintaining efficiency improvements. By improving efficiency by 7% to 12%, the $65,000 annual cost savings could be achieved. Consistent and accurate data collection (process observations and time studies) over several days was necessary. Other lessons learned related to the need to involve all team members (operators, managers, and maintenance employees) in the decision and improvement process to foster buy-in.

References

1. Henderson, K.M. and Evans, J.R. (2000). Successful implementation of six sigma: Benchmarking: General Electric Company. *Benchmarking: An International Journal*, 7(4): 260–281.
2. Breyfogle, F.W., Cupello, J.M., and Meadows, B., (2000). *Managing Six Sigma: A Practical Guide to Understanding, Assessing, and Implementing the Strategy That Yields Bottom-Line Success*. New York: Wiley-Interscience.
3. Pyzdek, T. and Keller, P. (2003). *The Six Sigma Handbook*. New York: McGraw-Hill.
4. Tompkins, J. and White, J. (1984). *Facilities Planning*. New York: John Wiley & Sons.
5. Park, S.H. (2002). Six sigma for productivity improvement: Korean business corporations. *Productivity Journal*, 43(2): 173-183.
6. Pande, P. and Holpp, L. (2002). *What Is Six Sigma*. New York: McGraw-Hill.
7. Yang, K. (2005). Design for six sigma and value creation. *International Journal of Six Sigma and Competitive Advantage*, 1(4): 355–368.
8. Kwak, Y.H. and Anbari, F.T., (2006). *Benefits, Obstacles, and Future of Six Sigma Approach*. New York: Elsevier Science.
9. McAdam, R. and Lafferty, B. (2004). A multilevel case study critique of six sigma: Statistical control or strategic change? *International Journal of Operations and Production Management*, 24(5): 530–549.

16

Case Study 5: Healthcare

16.1 Introduction

In manufacturing, Lean Six Sigma has been used extensively with great success. The application of Lean Six Sigma to the healthcare field is in its early stages and, hence, has not been fully explored. This study investigated the use of Lean Six Sigma with the goal of improving the renal implant process and demonstrating the positive impact of Lean Six Sigma on the healthcare industry. The objective of the research was to improve the process for renal transplants at The University of Toledo Medical Center utilizing Lean Six Sigma. This included aligning and optimizing processes and the removal of process-generated defects and errors.

The goal of the study was to analyze and improve the preoperational processes for patients undergoing kidney transplants. The Lean Six Sigma methodology was applied with the goals of: (1) optimizing cycle times, (2) enhancing customer satisfaction, (3) improving efficiencies, (4) reducing costs, (5) streamlining administrative processes, (6) eliminating errors, and (7) improving protocol execution and effectiveness.

The primary metric for this project was the number of days required from the date a patient was referred to The University of Toledo Medical Center for a kidney transplant to the date that the patient was "cleared" by the hospital staff as a suitable candidate for such atransplant. The primary need of this research was centered on an increase in the number of transplant patients "wait-listed" for transplants and the number of transplants performed per year due to an increased service area for the Medical Center. The waiting list is currently nearly 500 patients and a reduced cycle time will save lives.

For more than 30 years, The University of Toledo Medical Center has offered adult and pediatric kidney transplantation as one of the treatment options for end-stage renal disease. Since the first kidney transplant operation was performed at The University of Toledo Medical Center in 1972, more than 1,500 kidney transplant operations have been performed, with an average patient survival rate of 98% and a graft survival rate of 94%. The program relies on advanced surgical techniques including laparoscopic kidney donation, improved antirejection medications, and high-quality patient care to make it one of the most successful in the country.

16.1.1 Background and Terminology

There are a number of steps that patients must complete prior to receiving a kidney transplant. Generally, someone in need of a kidney transplant must be referred to a transplant center, complete the required labs and tests, and be considered a healthy candidate. The labs and tests are usually similar from one transplant center to the next and between patients. The labs include tuberculosis tests, dental clearance, colonoscopy, chest x-rays, EKG tests, stool samples, blood work, mammograms, pap smears, and diabetes tests. Once the patient completes the requirements, the information is reviewed by a committee to determine if the patient is a good candidate. The patient is then allowed to receive a kidney; this is called being "listed" or placed on the waiting list.

Often, the time required for a patient to complete all the health screenings for a kidney transplant can be nine months. In addition to this time, it may require another two years for the patient to receive a kidney transplant after being listed. It is difficult to decrease the waiting period between being listed and receiving a kidney because this time is dependent on the number of kidneys available and the number of kidneys needed; however, improvements can be identified for reducing the time required for a patient to become listed. The Lean Six Sigma team was comprised of two surgeons, three transplant coordinators/registered nurses, a psychologist, a financial advisor, and two industrial engineers. The methodology for this project was adapted from a plan developed from the book, *Improving HealthCare Quality and Cost with Six Sigma* [1] *and Partnering With Your Transplant Team, The Patients Guide to Transplantation* [2], developed by the US Department of Health and Human Services.

16.2 Application of Lean DMAIC Process

16.2.1 Define

The objective of this project was to improve the preoperational processes for patients undergoing kidney transplants by reducing the cycle time to 180 days or less. The primary metric for this project was the average total preoperational process time, measured in days, for patients undergoing kidney transplants as calculated from patient records. As discussed later in the analysis phase, the process was baselined at an average preoperational process time of 227 days for the preceding two-year period. To ensure that the process was patient-centered, concepts from the book, *What Every Patient Needs to Know* [3], were applied. These concepts revolved around patient involvement and communication with the care team.

16.2.2 Measure

In this phase of the project the key metrics were identified, the data collection process was developed and executed, the performance and variation of the current system were baselined, and the initial Sigma level was calculated. Included in this phase was the development of a supplier, input, process, output, and customer (SIPOC) diagram and process map. The SIPOC diagram can be seen as a high-level process map. It helps the improvement team to understand the purpose and scope of a process clearly. It is a starting point in identifying the voice of the customer (VOC). It provides initial insight into the vital inputs (or X variables) of a process [$Y = f(X)$] that have significant impact on critical outputs (or Y variables). It also becomes a primary input to detailed process map construction. Figures 16.1 and 16.2 are the SIPOC diagram and process flow map that were created to describe the process.

The output variable analyzed for this process was the time required from the time a patient was referred by his or her doctor to The University of Toledo Medical Center to the time the patient was evaluated by the medical staff and cleared as a suitable candidate for a transplant and a place on the wait list. The Six Sigma team conducted a brainstorming session using a fishbone diagram as shown in Figure 16.3 to identify causes of variation in the system.

The input variables (potential vital Xs) included: (1) lack of real-time patient data, (2) varying data collection procedures by doctors and nurses, (3) scheduling conflicts between staff and patients, (4) lack of communication between staff, (5) staff availability, (6) patient delays and cancellations, and (7) lack of funds from the patient. Data were collected for this project for the previous two years utilizing a combination patient record review, database

Suppliers	Inputs	Processes	Outputs	Customers
>Medical supply companies	>Patients	See Process Flow Diagram in Figure 16.2	>Implanted kidney	>Patients seeking transplants
>Insurance companies	>Donated organs		>Payment	>Family and friends of patients
	>Medical staff			>Insurance companies
				>Medicare

FIGURE 16.1
SIPOC diagram.

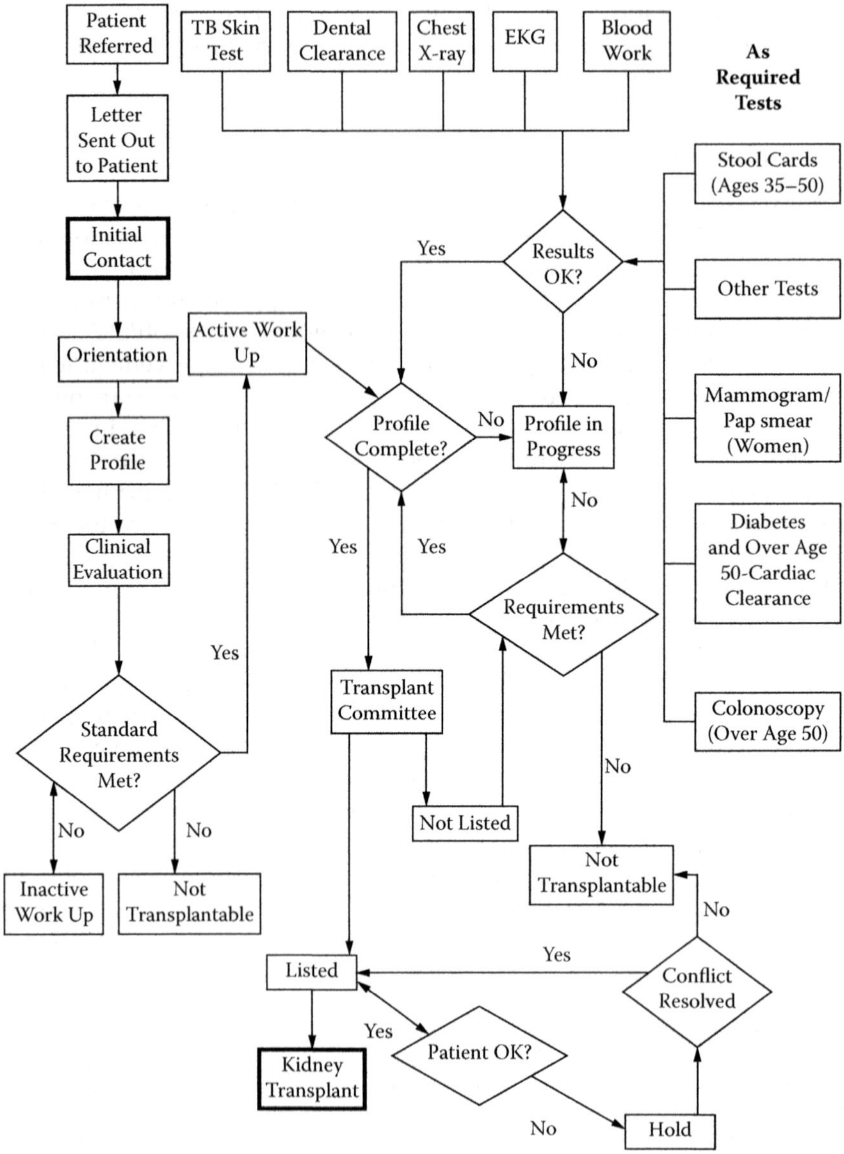

FIGURE 16.2
Process flow diagram.

review, and real-time data collection from current patients. The data were collected by the cross-functional team using standardized forms. Meaningful Six Sigma metrics were established based on the data and goal of the project. These metrics included the process cycle time, Sigma quality level, process capability, defects per million opportunities (DPMO), and yield.

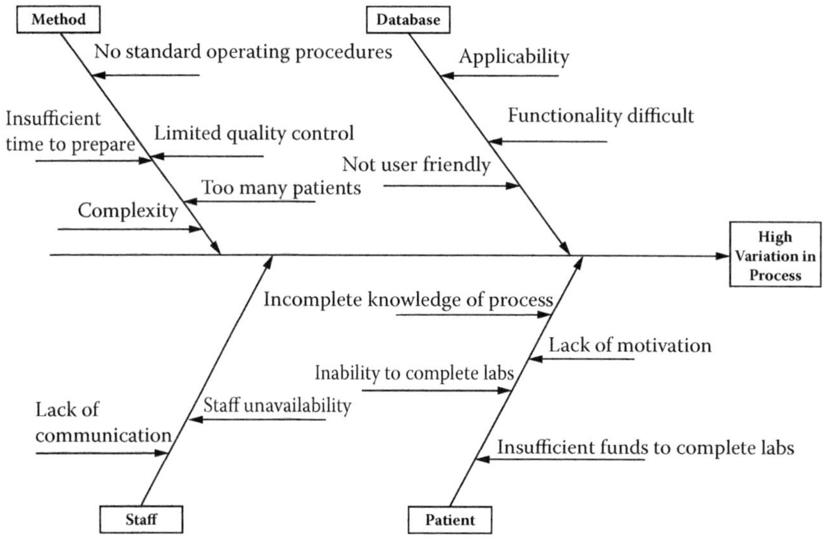

FIGURE 16.3
Fishbone diagram.

The Medical Center had previously established a goal of six months (180 days) for the completion of the evaluation process for each patient. This goal was established based on the minimum time required for the various tests, and patient delays, and the critical nature of transplants to save lives. The current mean process cycle time was 227 days with a standard deviation of over 45 days. Based on a sampling of 509 patients, 433 patients (85.1%) required longer than 180 days for the evaluation process. This translates into less than 15% of all patient evaluations achieving the 180-day goal. The 14.9% acceptance rate translates into 851,000 defects per million opportunities or a 0.46 Sigma level for the current process.

16.2.3 Analyze

Various Six Sigma tools were applied to understand process capability and identify the vital Xs or the root causes of process delays requiring longer than 180 days to complete. The tools and data analysis methods applied included: (1) process capability analysis, (2) confidence and prediction intervals, (3) Pareto analysis, (4) five Why analysis, (5) root cause analysis, (6) regression analysis, (7) failure modes effect analysis, (8) work sampling, and (9) value stream analysis.

Before identifying and analyzing root causes, a baseline for the current process performance was established and a process capability analysis was performed. Table 16.1 summarizes the findings from a review of over 500 patient records over the past five years.

TABLE 16.1

Process Baseline Measurement

	Time Between Referred and Letter	Time Between Letter and Orientation	Time Between Orientation and Clinical Evaluation	Time Between Clinical Evaluation and Transplant Committee	Time Between Transplant Committee and Listed	Total Process Time
Average days	5	40	68	102	13	227
St. dev (days)	6	22	27	82	15	45.6
Average months	0.16	1.34	2.27	3.39	0.42	7.58
St. dev (months)	0.19	0.72	0.89	2.74	0.50	1.52
# of samples	91	81	97	136	104	509

As mentioned in the Define phase, the current mean time was 227 days with a standard deviation of over 45 days. To validate the normality assumption for this dataset a histogram was created to display the central tendency. Based on a sampling of 509 patients, 433 patients (85.1%) required longer than 180 days for the evaluation process, meaning less than 15% of all patient evaluations achieved the 180-day goal. As shown in Table 16.2, the high process average and standard deviation for the time between when the patient receives a letter and orientation is because orientation is offered once a month. The clinical evaluations are offered two days per week and are scheduled for months, thus contributing to the high average and standard deviation for the time between orientation and the clinical evaluation. The time between the clinical evaluation and transplant committee has such a high average and standard deviation because the patient has to complete all

TABLE 16.2

Confidence Interval

95% Confidence (Days)	Time Between Referred and Letter	Time Between Letter and Orientation	Time Between Orientation and Clinical Evaluation	Time Between Clinical Evaluation and Transplant Committee	Time Between Transplant Committee and Listed	Total Process Time
z-Value	1.96	1.96	1.96	1.96	1.96	1.96
Interval	1.23	4.79	5.37	13.78	2.88	3.96
Upper bound	6	45	73	116	16	231
Average	5	40	68	102	13	227
Lower bound	4	35	63	88	10	223

of the required labs and tests. The number and specific tests required are based on the individual. The average and standard deviation for the time between the transplant committee and being listed is higher than expected. This is due to the workload of the coordinators. From a process capability standpoint, the process is not capable of performing within specifications in the current state as calculated with a *Cpk* value of –0.34. The negative *Cpk* value highly suggests the process is well on the high side of processing under 180 days.

A confidence interval (Table 16.2) was calculated to show the range of the true mean. This is beneficial to the patient and staff to give them an idea for the true means of the process times. This baseline calculation also allowed for a statistical analysis of improvements. As displayed in the table, at a 95% confidence level, the true process cycle time is between 223 and 231 days.

These tools were applied to identify the assignable causes leading to the gap between the current performance and the goal. Numerous reasons such as surgeon delays, coordinator delays, patient delays/cancellations, communication, and patient failure to complete lab tests contributed to the process delays over 180 days. After developing the process map and collecting data for 116 active patients, Pareto, and Five Why analysis indicated patient delays in completing lab tests were the top reasons for failures. To understand which stage was restraining patients from being listed, as displayed in Figure 16.4, a Pareto chart was created to highlight which labs were incomplete.

The chart in the figure indicates that the following tests were incomplete: 57% of patients required a cardiac clearance, 40% of patients required dental clearance, 40% of patients required a CT or CTA scan, and 35% of patients required a colonoscopy.

A multiple regression analysis was conducted to identify the significant variables that influence the overall process time using stepwise regression to evaluate the data. The following variables were evaluated: TB skin test, dental clearance, cardiac clearance, colonoscopy, mammogram/Pap smear, CT/CTA scan, other tests, age, gender, distance, and whether the patient was listed at another facility for a transplant.

Table 16.3 displays the results of the regression analysis. After three steps, 34% (adjusted $r2 = 0.341$) of the variation in the total process time was explained by whether the patient required a CT test, was listed elsewhere, or required a TB exam. This relatively low adjusted $r2$ value indicates that the model is not a strong predictor of total process time, but does offer some insights into the process, both in the variables that were included and those that were not included.

A failure mode effects analysis (FMEA) was conducted. The Six Sigma team analyzed the data and system to determine the failure modes, potential effects, severity, occurrences, and detection. The key findings from the study are listed below and the actions taken are listed in the Improve section. A work study was completed to evaluate which tasks occupy most of the transplant coordinator's work day (Figure 16.5). The majority of the coordinator's

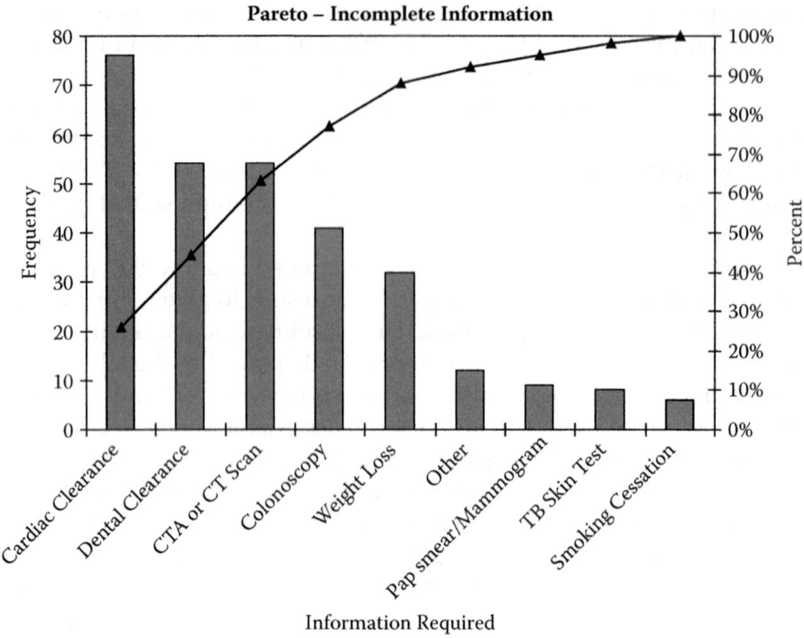

FIGURE 16.4
Pareto analysis.

TABLE 16.3
Multiple Regression Results

Number of Steps	Variable	Step Constant	Days Added to Step Constant	Total List Time	T-Value	P-Value	S-Value	R-Sq Value	R-Sq (Adj)
1		61					70.3	28.18	25.6
	CT/CTA scan		87	148	3.31	0.003			
2		79					68.8	33.61	28.7
	CT/CTA scan		79		3.02	0.005			
	Listed elsewhere		–40	118	–1.49	0.149			
3		135					66.1	40.95	34.1
	CT/CTA scan		94		3.56	0.001			
	Listed elsewhere		–57		–2.09	0.046			
	TB test		–67	105	–1.80	0.084			

time was spent preparing and reviewing the patients' files which consisted of organizing information, filling out and getting required information, updating the patient's medical history and information, reviewing the patient's results, and sending patients the necessary information. The other activities consisted of sending and receiving e-mails, checking mail, organizing

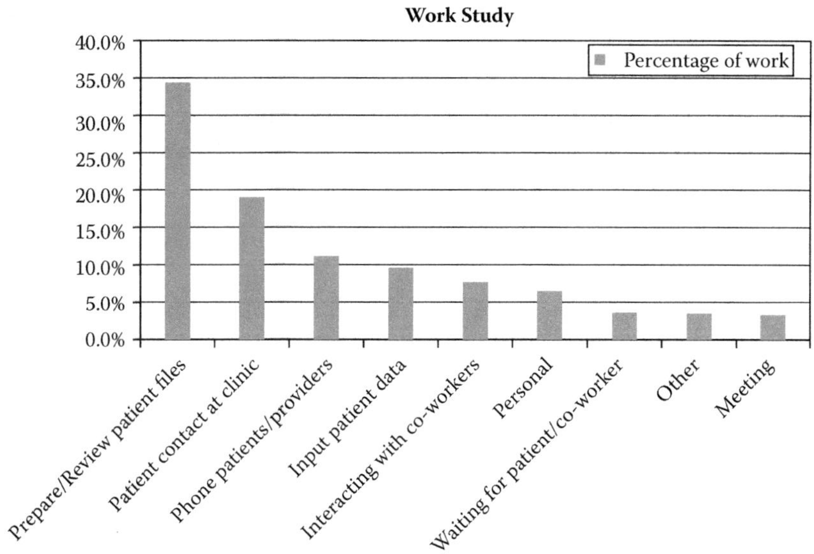

FIGURE 16.5
Work study.

and scheduling, reviewing meeting notes, and other miscellaneous activities. Figure 16.5 shows that 34% of the coordinators' time is spent preparing patients' files, meaning one-third of their time is preparing to see patients.

Upon analyzing the work study, the coordinator had idle time for roughly an hour a week waiting for other transplant members to complete their evaluations with the patient. This is non-value-added time and keeps the coordinator from doing more important tasks. Data entry should be minimal for the coordinator. Figure 16.6 displays the percentage of value-added tasks and non-value-added tasks of the coordinator. Value-added tasks were defined as patient or provider contact (30% of activities). Non-value-added

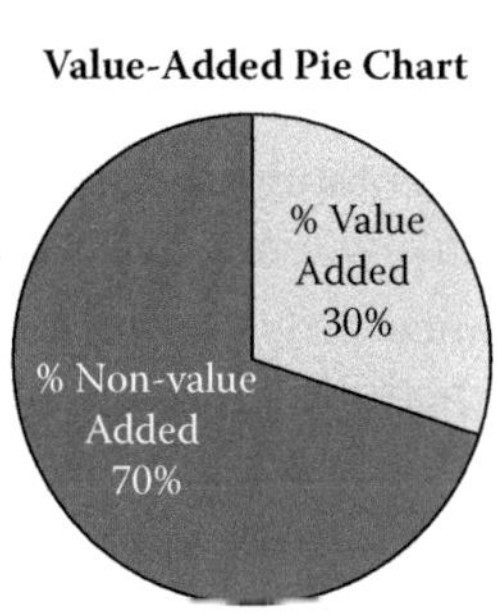

FIGURE 16.6
Value-added pie chart.

activities were defined as waiting for patients/staff, data input, and data review/preparation (70% of activities). The main focus of the coordinator should be working with the patients and coordinating labs and tests for the patient, so the patient can be presented as soon as possible. Tasks that should be focused on to be reduced are: waiting for patient/coworker, inputting patient data, and preparing/reviewing patient files.

16.2.4 Improve

After identifying and analyzing the potential vital Xs, the team created a list of improvement initiatives to address each issue. Meetings and focus sessions were held with the patients, surgeons, nurses, transplant coordinators, and other staff to review the results and develop the following list of improvements:

- Creation and implementation of a patient database to provide information and to generate custom reports (incomplete tests, status).
- Tracking system in database to monitor performance to make sure targets are met based on timelines (tests, process performance, communication between staff and patient).
- Consider using a video for orientation instead of presentation (mailed or web-based). If used, consider not starting any paperwork until patient is signed up for evaluation.
- Consider using a presheet process (overview of process given at orientation).
- Modification to the clinical schedule to improve efficiency of medical staff, and reduce patient cycle time (establish evaluation so coordinator only has to see the patient at the end). This would allow coordinator to spend less time at clinical evaluation and more time to contact patients (contact person to break barriers). Consider moving social worker into different office to conduct evaluation.
- At end of clinical evaluation give a simplified handout/contract describing process and expectations with timeline.
- Time it takes between orientation and the clinical evaluation is high. Consider offering clinical evaluations an additional day of the week.
- Develop dental assistance program to identify potential doctors/costs and assist with scheduling for patient.
- Continuous monitoring system based off the database. Send report out to patient to verify information at a determined interval (two weeks, four months).
- Develop standard operating procedures for process, to ensure uniformity and consistent work.

The transplant team has decided to pursue a database. The database will have the ability to track important data and performance measures. To improve the clinical evaluation process the team modified the patient schedule by moving the financial officer and social worker to see patients after the doctor and coordinator are finished with them, which may be done in a different room. The team does not currently have enough staff to offer clinical evaluations an additional day, but they have agreed to add another patient to each day that they do offer evaluations. They are also working with different dental offices to allow for decreased dental costs to patients who cannot afford dental care.

The transplant team implemented a "solution center." This created a position to handle incoming calls about the transplant process. This center required employees who were able to be on the database to look up information and answer common questions. If they could not answer a question they were to direct the patient to the appropriate staff member. This would allow coordinators time to focus on helping patients become listed faster. The "solution center" made the process more personal to the patient because the patient would only have to call one number to have his or her questions answered. This helped to avoid the frustration of patients calling many different places and not getting an answer. The "solution center" also served as a "barrier breaker" and further personalized medical services.

As displayed in Table 16.4, a total process cycle time reduction of 63 days (27.8% of the current cycle time of 227 days reduced to 163 days) was identified. The improvements were implemented and the results were shared daily with the medical staff and administration. After meeting with all stakeholders, a standardized process was created and documented. The goals of the standardized process were to "mistake-proof" and streamline the process from a patient viewpoint. The process was tracked and the process cycle time and standard deviation were measured for each patient for a one-year time period. The mean cycle time was reduced by 28.2% (64 days) to 163 days and the standard deviation was reduced from 45 days to 27 days.

To validate the process improvements, a process capability study was completed on the process cycle time for the improved system. Also, hypothesis tests were conducted on the difference between the means and variances for the previous process and the new process. From a process capability standpoint, the new process has significantly improved over the previous process. The new process is much more capable of performing within specifications (cycle times within 180 days) as calculated with a *Cpk* value of 0.21, up from –0.34 of the previous system, although work is still needed to raise the *Cpk* to a value greater than 1.0. Assuming a normal distribution with a mean of 163 and standard deviation of 27 days, 73.6% of all cycle times will be within in target process time of 180 days or less. To validate this normality assumption, a histogram was created for the processing times after the improvement initiatives were implemented. Figure 16.7 displays this histogram and based on the expected bell-shaped curve, the normality assumption appears valid.

TABLE 16.4

Estimated Improvements and Initiatives

	Process Issue	Proposed Change	Quantified Estimated TPTR	Qualified Improvement
1	Lack of communication and organization	Use a database	1% (2 days)	Improved staff efficiency and communication
2	No performance tracker	Use a database	0%	Improved staff efficiency and communication
3	Frequency of orientation	Develop a video	7% (16 days)	Improved customer satisfaction
4	Unclear process expectations	Use pre-sheet process overview	1% (2 days)	Improved customer satisfaction
5	Idle time at clinic	Change clinical schedule	0%	Improved use of staff time
6	Unclear requirements	Simplify contract/handout	1% (2 days)	Improved customer satisfaction
7	Frequency of clinical evaluation	Offer clinic more	8% (18 days)	Improved customer satisfaction
8	Dental lab incompleteness	Offer dental assitance program	2% (5 days)	Improved customer satisfaction
9	No patient continuous monitoring system	Use a database	0%	Improved patient feedback
10	No current standard operating procedure	Create standard operating procedures	0%	Improved staff efficiency and communication

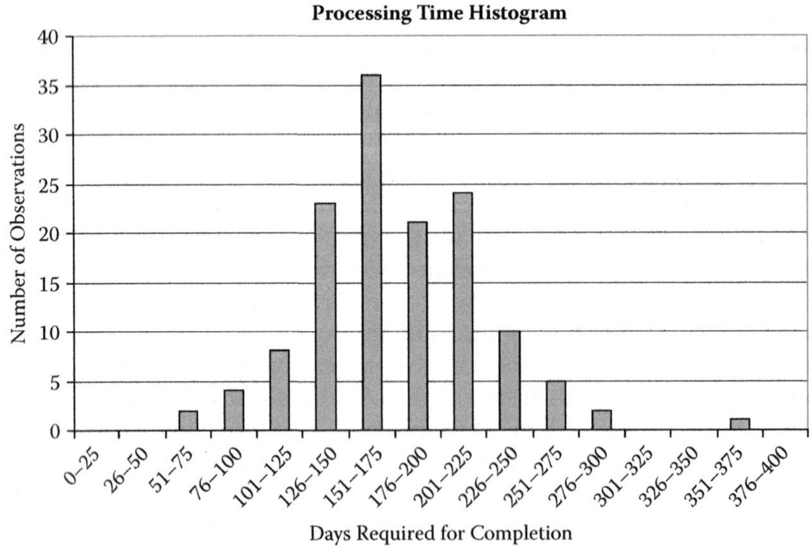

FIGURE 16.7
Histogram of completion times for the improved process.

Training courses and standard operating procedures were developed for all processes. As a result of the process improvements, process cycle time failures (defined as greater than 180 days) were reduced by 68.9% from 851,000 to 264,000 defects per million opportunities. This reflects an annual savings of $22,000 from reduced administrative costs and photocopies.

16.2.5 Control

To ensure the process performs within acceptable limits and to continue to drive down the cycle time, the performance is monitored on an ongoing basis. To achieve "control" status, x and r bar charts are utilized. In addition, weekly meetings with the key staff are held to discuss performance and to generate new improvement solutions. These meetings involve a review of process times, patient surveys, and adherence to standard operating procedures.

References

1. Trusko, B.E., Pexton, C., Harrington, J.H., and Gupta, P. (2007). *Improving HealthCare Quality and Cost with Six Sigma*. Trenton, NJ: Pearson Education.
2. US Department of Health and Human Services. (2005). *Partnering with Your Transplant Team: The Patients Guide to Transplantation*. Department of Health and Human Services.
3. Pfeiffer, A., Trinkle, B., and Mock, L.K. (2007). *What Every Patient Needs to Know*. United Network Sharing.

17

Case Study 6: Manufacturing—Small and Medium-Sized Operations

17.1 Introduction

The company studied was a medium-sized manufacturer of riding lawn mowers based in the midwestern United States. The top-selling product for the company was a riding lawn mower unit and the company employed 85 people and operated in one production shift. The company made several engineering design changes and began to experience quality failures in the drive chain of the lawn mowers. The purposes of the study were to understand the cause of the defects, quantify the amount of defects, identify the cause, and eliminate the defects.

17.2 Application of Lean DMAIC Process

17.2.1 Define

For the first phase of the project, the key quality issues from the customers' perspective were identified. Based on customer complaints and engineering sales feedback, the drive chain of the lawnmower would become jammed and in some cases, dislodge from the assembly housing (Figure 17.1). The Lean Six Sigma team conducted a Five Why analysis, and studied the assembly process and end units to identify that the machined diameter of the drive chain and the assembly housing were the primary quality issues that led to breakdown or failure.

17.2.2 Measure

A process map and a SIPOC (supplier, input, process, output, and customer) diagram were created for the drive chain assembly as displayed in Figures 17.2 and 17.3. These tools allowed the Lean Six Sigma team to identify the major quality issues clearly.

FIGURE 17.1
Drive chain assembly.

The primary tool used in the measurement phase included a repeatability and reproducibility study to ensure that the measurement system was accurate and precise. These studies measure the actual variation in parts and the measurement error due to operators. The sample size of 20 was selected and two readings were taken on each sample, for a total of 40 readings using a micrometer. From the results of study, repeatability and reproducibility were calculated as 12.6% and 0.0% error, respectively; these results indicate that the gauge is correct as operator error (variation is very low and total error is less than 30%).

17.2.3 Analyze

A process capability analysis was first completed to confirm that the process was capable of meeting the customer's requirements. The process capability ratio was calculated to be 1.35; the fact that it was greater than 1.00 indicated that the process was capable of performing within the customer requirements. A fishbone diagram was created to identify sources of variation (Figure 17.4). After further study of three operators, operator variance appeared to be a large source of error based on a statistical sampling of 60 parts from the same machine (20 parts per operator). A *t*-test was conducted to confirm that a statistically significant difference existed between the operators at the 0.05 level of significance (Table 17.1). Based on the fishbone diagram and *t*-test results, the Lean Six Sigma team focused improvements on operator skill and training.

17.2.4 Improve

The improvement phase focused on applying the information gathered in the analysis phase by conducting a design of experiment (DOE) to identify the optimal conditions to ensure consistent quality within the customer

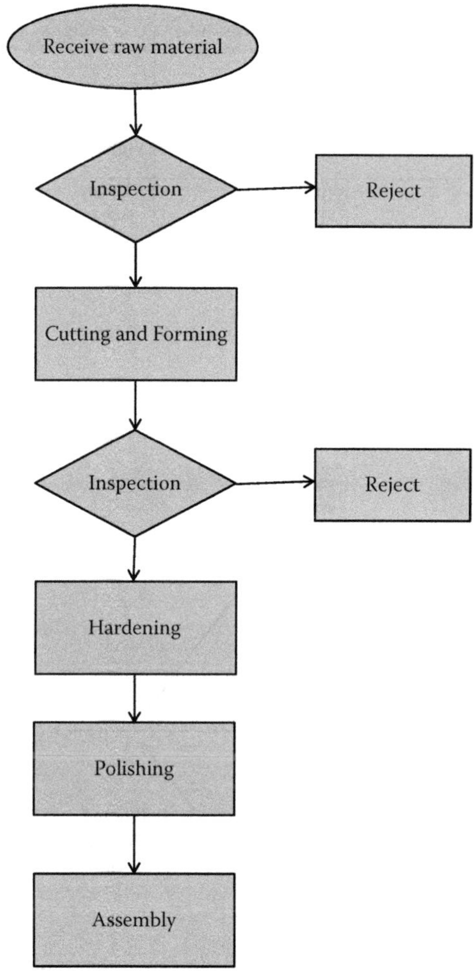

FIGURE 17.2
Process flowchart.

specifications. The initial conditions were discovered based on the results of the two-sample *t*-tests discussed in the previous section. The purpose of the DOE was to optimize the parameters of the drive chain and assembly housing. A 2 × 2 DOE experience was designed. Plots were created to study the main effects and interaction between the vital few factors (drive chain and housing). The main effect plot indicated that the drive chain is a minor factor and the housing unit a major factor. The Lean Six Sigma team worked closely with the design engineer, manufacturing engineer, and supplier to redesign the housing unit to prevent failures. The operators who machined the housing were also given detailed training to standardize procedures and eliminate defects.

Suppliers	Inputs	Processes	Outputs	Customers
>Metal	>Customer acceptance data	>Process capability and process flows	>Improved yields	>Customer satisfaction
>Chain assembly			>Timely customer payment	>Lead time

FIGURE 17.3
SIPOC diagram.

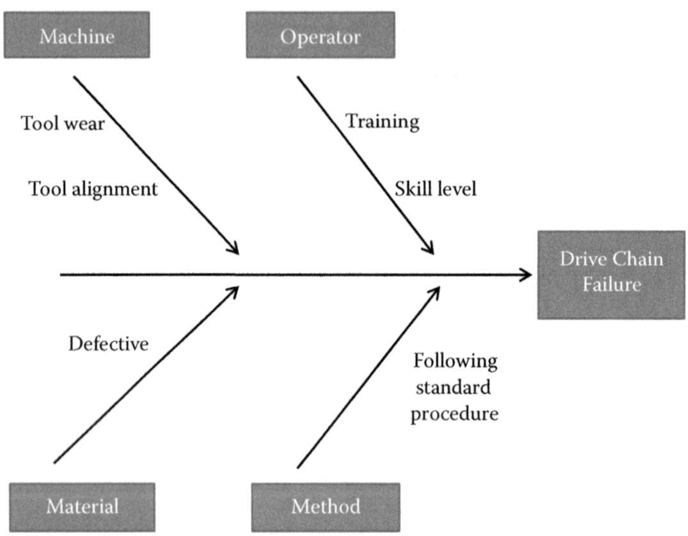

FIGURE 17.4
Fishbone diagram.

TABLE 17.1

Two Sample *t*-Tests for Machine Diameter (Inches) and Operator

Operator	Sample Size (N)	Mean	SD	SD Error
1	50	8.164	0.015	0.0018
2	50	8.091	0.008	0.0012

17.2.5 Control

To ensure that the drive chain and housing are machined within specifications, the performance was monitored on an ongoing basis. To achieve "control" status, x and r bar charts were utilized. In addition, weekly meetings with the key staff were held to discuss performance and to generate new improvement solutions. In addition, extensive operator training was conducted and a training manual was developed to ensure that all operators continue to machine parts in a similar manner.

18

Case Study 7: Manufacturing—Automotive Supplier

18.1 Introduction

The company studied was a first-tier automotive supplier operating in Ohio and Indiana, USA. The company employed 650 people working in three shifts at three production plants. Each of the three plants had full maintenance, engineering, and finance staff. One plant produced raw material and two plants produced parts from the material. The company produced fiberglass sheet molding compound (SMC) exterior parts for Ford and General Motors. The parts were produced using injection molding and included fenders, truck lids, hood lids, and other exterior trim (over 2,500,000 parts per year between the two production plants). At the time of the study a major quality defect called "bubbling" was appearing on the outside of the parts, apparently random in terms of which parts and location. The bubbling defects were cause for rejection by Ford and GM and the parts could not be reworked. The purposes of the study were to identify the cause and eliminate the bubbling defects while streamlining and standardizing the production processes.

18.2 Application of Lean DMAIC Process

18.2.1 Define

The first steps of the Lean Six Sigma process were to form the team, clearly understand the issues, and define the problem. The team consisted of a manufacturing engineer, a materials specialist, a quality control technician, a line worker, supervisor, and an engineering intern. As mentioned, the primary defect related to bubbling on the surface of injected molded exterior automotive parts. The defects started to appear in late June and early July, apparently randomized based on machine, part, and operator. During the

June and July time period, approximately 15 to 20% of the parts produced at the two production facilities had these defects of varying degrees. While defining the quality problem, the team also observed large amounts of work in process waiting between machine centers. Based on these issues goals of the Lean Six Sigma project were:

1. Eliminate the bubbling defect in all injection molded parts.
2. Reduce work in process by 50%.
3. Reduce average product cycle time by 15%.

18.2.2 Measure

To first better understand the process, the team created a process flowchart for the injection-molded parts after the raw material arrived at the facility (Figure 18.1) To better understand processing times for Lean improvements, the team, led by the industrial engineer, conducted a time study and downtime analysis for the press operation process (these results are discussed in the following analysis section).

18.2.3 Analyze

The analysis phase involved studying the data collected during the measurement phase to eliminate the bubbling defect and make the process Leaner by reducing waste. The first action for the team was to correct the process by eliminating the defect. Once the defect was eliminated, the process could be made Leaner and standard operating procedures developed.

18.2.3.1 Quality Improvement

To begin, the team held a meeting to conduct a failure modes effects analysis (FMEA) and a five Why analysis. Based on the analyses, the team identified five potential causes of defects:

1. Operator error (loading, unloading, or machine cycling errors)
2. Machine variation (temperature or pressure)
3. Raw material defects (nonobservable at visual inspection)

The team analyzed the data further to pinpoint the exact cause or causes of the defects based on the FMEA. An analysis of variance (ANOVA) table was created to understand the variance and quality issues associated with certain presses (machines), operators, materials, and product lines. To conduct the ANOVA, separate studies were conducted to examine differences in the number of defects between machines, operators, raw material, and product lines over a three-week period. Differences between machines, operators,

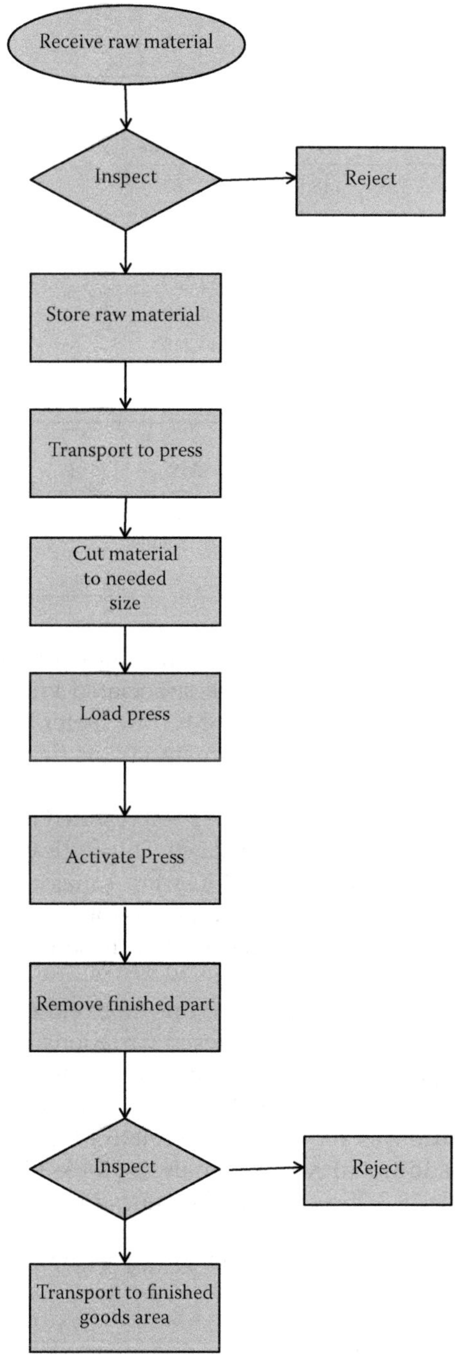

FIGURE 18.1
Process flowchart.

TABLE 18.1

ANOVA for Causes of Defects

Summary

Groups	Count	Sum	Average	Variance
SMC 1	21	395	18.810	12.562
SMC 2	21	33	1.571	1.357
SMC 3	21	398	18.952	5.648
SMC 4	21	423	20.143	12.129
SMC 5	21	61	2.905	4.890
SMC 6	21	1310	62.381	13.248

ANOVA

Source of Variation	*SS*	*df*	*MS*	*F*	*P*-value	*F* crit
Between SMC	50961.968	5.000	10192.394	1227.178	0.000	2.290
Within days	996.667	120.000	8.306			
Total	**51958.635**	**125**				

and product lines were not statistically significantly different at the 95% confidence level, but the number of defects associated with the raw materials was statistically different for different SMC raw material lots. Table 18.1 displays the ANOVA analysis for the raw material lots (labeled SMC 1 through SMC 6).

As displayed in Table 18.1, the ANOVA *f*-test resulted in a calculated *f*-value of 1227.178 versus the critical *f*-value of 2.290. This indicated a very high probability that raw material lots for the SMC were a major cause of the defects. The calculated *f*-values for the machines, operators, and product lines were below the 2.290 critical value, indicating that these parameters were not a statistically significant cause of defects.

The team analyzed the raw materials runs more closely, working with the operators to understand SMC raw material variations better. The team discovered that during periods of high temperature and high humidity in the plant and raw material storage area, the defects rate was significantly higher when the temperature was above 74°F and relative humidity was above 55% as displayed in an additional ANOVA analysis (Table 18.2).

As shown in Table 18.2, the calculated *f*-values for differences in the number of defects at higher temperatures and relative humidity were well above the critical *f*-value of 2.48 and 3.26 at 213.59 and 320.97.

18.2.3.2 Lean Process Design

After the primary cause of defects was determined and potential solutions were identified, the team studied the overall injection molding process to

TABLE 18.2

ANOVA for Temperature and Humidity

ANOVA						
Source of Variation	*SS*	*df*	*MS*	*F*	*P*-value	*F* crit
Sample	2946.76	5	589.35	213.59	0.00	2.48
Columns	1771.26	2	885.63	320.97	0.00	3.26
Interaction	979.63	10	97.96	35.50	0.00	2.11
Within	99.33	36	2.76			
Total	**5796.98**	**53**				

TABLE 18.3

Time Study Results

Minimum cycle	1.83 min
Average cycle	2.31 min
Standard deviation	1.08 min
Runtime	469.10 min
Downtime	97.74 min
% Uptime	79.16%

make it Leaner. Table 18.3 displays the results of the time study conducted on the operation.

As displayed in Table 18.3, the current process cycle time (standard time) was 1.83 minutes and the average cycle (observed) was 2.31 minutes. The average observed cycle time was 26.2% over the standard cycle time of 1.83 minutes. Next, the team analyzed the downtime associated with the injection molding presses. Table 18.4 displays the downtime analysis causes and Figure 18.2 displays the Pareto analysis for the causes.

TABLE 18.4

Downtime Analysis Results

Machine: Press#1 (3 Operators)							
					Accumulated Pieces Lost		
Reason for Downtime	**Time Lost (Min)**	**% of Total**	**Frequency**	**Pieces Lost**	**Shift**	**Day**	**Year**
Unauthorized break	32.86	46.4	1	18	38	115	29,836
Machine loading delay	15.37	21.7	24	8	45	74	35,093
Raw material change	14.14	20.0	2	7	41	68	32,285
Unscheduled maintenance	3.59	5.1	3	2	11	17	8,197
Employee delay/talking	2.95	4.2	2	1	9	14	6,735
Machine unloading delay	1.92	2.7	21	1	6	9	4,384
Totals	**70.83**			**37**	**149**	**297**	**116,529**

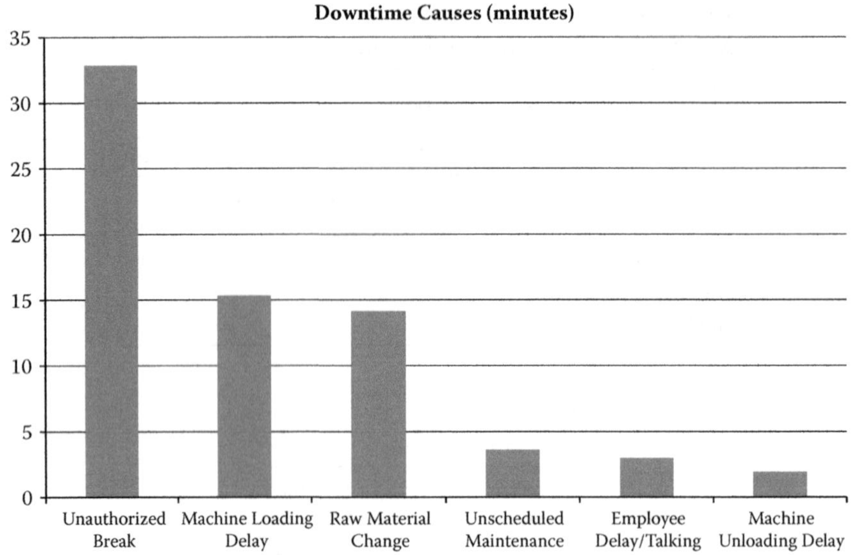

FIGURE 18.2
Pareto analysis for downtime causes.

The press was down for approximately 70.8 minutes of the operational time of 266 minutes studied; this translated into over 116,000 lost final parts per year for the one press studied. The plant utilized 22 presses, so the facility was losing up to 2.5 million parts per year due to downtime and not meeting the standard cycle time. As shown in Figure 18.2, the primary causes of the losses were unauthorized breaks, machine loading delays, and raw material changeovers. Solutions to these issues are discussed in the following Improve section.

18.2.4 Improve

18.2.4.1 Quality Improvements

The plant and raw material storage area were not temperature or humidity controlled. When higher temperatures (over 74°F) and high humidity (over 55%) were present, higher bubbling defects appeared as displayed earlier in Table 18.2. To test the hypothesis, the team stored the raw material in a temperature-controlled office overnight (65°F and 35% relative humidity) during a period of high heat and humidity in the plant (over 85°F and 65% relative humidity). The next morning the team processed parts from the raw material held in the temperature-controlled environment and from the current material storage area. The results regarding the number of defects are displayed in Table 18.5.

TABLE 18.5

Defect Rates for Controlled Experiment

	Number of Parts Made	Number of Defects	% of Defects
65 Degrees/35% relative humidity	550	2	0.4
85 Degrees/65% relative humidity	548	48	8.8

As shown in Table 18.5, the defect rate was significantly higher for the material that was not in a temperature- and humidity-controlled environment (0.4% vs. 8.8%). As displayed in Table 18.5, the temperature and humidity were statistically significant and confirmed that the team had identified the root cause of the bubbling defect.

18.2.4.2 Lean Process Design Improvements

By redesigning, monitoring, and controlling the injection molding process by removing the avoidable delays, the team was able to reduce the cycle by 20.8% to the standard cycle time of 1.83 minutes. This was accomplished by using the results of the Pareto analysis to:

- Eliminate unauthorized operator breaks by assigning team leaders to monitor productivity rates for each press.
- Eliminate operator loading delays and ensure that the operator was ready to load the press with new raw material immediately after each machine cycle ended.
- Reduce the raw material change over time by moving prep work offline.
- Extensively train operators, team leaders, and supervisors on standard operating procedures and standard cycle times.

The company produced 2.5 million parts per year under the higher cycle time of 2.31 minutes, so the improved cycle time resulted in the generation of approximately 500,000 more parts and nearly $750,000 in additional revenue.

18.2.5 Control

The first action in the control phase involved adding temperature and humidity control to the raw material storage room for the SMC rolls. The room was kept at a constant temperature of 73°F and 50% relative humidity by installing an air-conditioning unit in the room. The cost of the unit and utility costs were marginal. Extensive training was provided to the operators, team leaders, and supervisors regarding the standard operating procedures, need for temperature controls, and new quality control practices. x Bar and r charts were used to track and monitor defects and hourly temperature/humidity readings were taken in the raw material storage area. As a result of the Lean Six Sigma efforts, the bubbling defect rated dropped to 0.02% and the average cycle time reduced to 1.89 minutes.

19

Case Study 8: Service—Emergency Plumber Dispatching

19.1 Introduction

The company studied was a 24-hour emergency and residential plumbing company. The company employed 22 plumbers and three dispatchers to take customer calls and schedule plumbing repairs or new work orders. The company was beginning to experience delays, higher costs, and decreased productivity from the plumbers. The purposes of the study were to understand the cause of the issues, quantify the issues, identify the cause, and improve productivity and customer service.

19.2 Application of Lean DMAIC Process

19.2.1 Define

For the first phase of the project, the key quality issues from the customers' perspective were identified and the productivity issues were recognized. The dispatchers were responsible for handling customer calls, quality issues, and scheduling plumbing jobs in a 24-hour per day, seven-day per week environment. A new dispatcher was added to the group to maintain and enhance customer satisfaction and scheduling, but improvements were not realized. The goal of the Lean Six Sigma project was to increase productivity by 10% on a year-over-year basis and increase customer service and response time of the plumbers by 20%.

19.2.2 Measure

The purpose for the measurement phase was to baseline the productivity rates for the two different plumber groups; group P1 who handled lower-level/lower-skill jobs and P2 who handled higher-level/higher-skills jobs.

Dispatcher Activities	Plumber Activities	Dispatcher Activities
Dispatcher Employee Number	Plumber Employee Number	Dispatcher Employee Number
Identification of Category P1 or P2	Plumber Employee Category (P1 or P2)	Plumber Employee Category (P1 or P2)
Identify and Dispatch Plumber	Travel Time to Customer Site	Enter Total Service Time and Cost
	Discussion with Customer to Pinpoint Issue	Close out Service Job/Accounting
	Cost Estimate to Customer	Clear Plumber for New Activities
→	Log Activities in Computer →	
	Supply Acquisition	
	Repair/Installation Work	
	Verification and Testing	
	Final Bill to Customer/Collect Payment	
	Log Activities in Computer	
	Travel to Next Customer Site	

FIGURE 19.1
Methodology to measure productivity and conduct time study.

To measure productivity, the Lean Six Sigma team used the ratio of the size of work to the effort involved to develop a comprehensive and combined metric. Figure 19.1 displays the methodology used to measure productivity and create the comprehensive ratio.

To determine the productivity rates for groups P1 and P2, a time study was conducted to measure the frequency of the defined activities and the average effort involved for performing that activity per day. The results for the time study are displayed in Table 19.1.

The team defined resource utilization as normal if it was within the range of 65% to 85%. Anything beyond this boundary was categorized as underutilized or overutilized. An overutilized resource would, in this instance, be someone requiring over eight hours on a given day, rather than the standard eight-hour shift, to finish the volume of work that he or she received. Having

TABLE 19.1
Daily Activities for Plumbers (in Minutes)

Activity	Customer A	Customer B	Customer C	Customer D	Customer E
Travel time to customer site	20	11	25	12	15
Discussion with customer to pinpoint issue	15	5	20	6	25
Cost estimate to customer	12	12	13	12	14
Log activities in computer	3	3	3	3	4
Supply acquisition	10	12	10	9	11
Repair/installation work	38	35	66	25	93
Verification and testing	8	7	5	6	8
Final bill to customer/collect payment	4	8	12	5	4
Log activities in computer	3	3	3	3	3
Break	0	0	30	0	0
Wait time before next job	0	0	0	12	N/A

gathered the frequency data of an activity's occurrence in a day and the average time of its performance, the team was able to compute the total effort invested by the team on each individual task. By examining team composition (defined by the number of P1 or P2 consultants and project manager) coupled with activity alignment per consultant type, the team determined the daily activity required for individual P1 and P2 plumbers.

19.2.3 Analyze

For the analysis stage, the Lean Six Sigma team reviewed the time study and process data to identify trends. From an initial review of the data, the team identified:

- High variance in utilization rates between plumbers and P1/P2 groups
- Periods of peak demand in mornings and early evenings for emergency calls

Figures 19.2 and 19.3 display the utilization rates for the plumbers employed by the company based on category. Figure 19.4 displays the time-of-day utilization rates for these plumber categories.

As displayed in Figures 19.2 and 19.3, plumber category P2 is consistently utilized and there was significant variation in plumber utilization rates for category P1 (ranging from 52% utilization to 91% utilization). The analysis of the overutilized and underutilized plumbers showed that the volume of service requests, customer service requests, and tracking plumber activity were the key drivers of utilization. As shown in Figure 19.4, there is more demand on the system in late mornings and early evenings (with peak

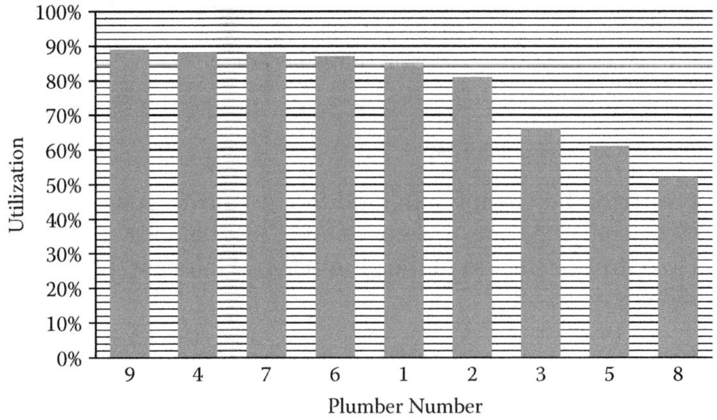

FIGURE 19.2
Plumber utilization rates by individual plumber (P1 category).

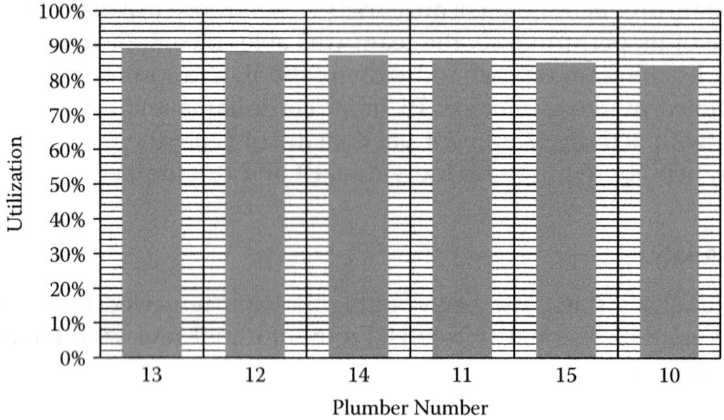

FIGURE 19.3
Plumber utilization rates by individual plumber (P2 category).

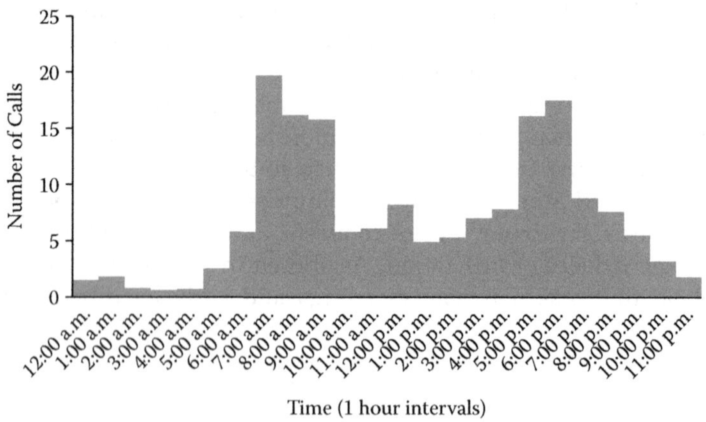

FIGURE 19.4
Plumber utilization rates by time of day.

demand occurring between 7 a.m. and 9 a.m. and then again from 5 p.m. to 7 p.m.). The challenge was to smooth and "line balance" the utilization of plumbers to optimize the use of the plumber and increase/reduce staffing to match customer demand.

19.2.4 Improve

Based on data examined in the analysis phase, the Lean Six Sigma team's first improvement approach was to track and monitor the performance and utilization of individual plumbers. This involved adding GPS tracking devices to the plumbers' service vehicles and creating a smartphone app to allow the plumbers to record service times and customer notes easily. This tracking and real-time information allowed for dispatchers to better schedule and utilize P1 and P2 category plumbers based on utilization rates and current locations based on GPS data. The second improvement approach was to hire and schedule the plumbers and dispatchers to match customer demand during peak periods. Three additional dispatchers were on duty between peak demand from 7 a.m. to 9 a.m. and then again from 5 p.m. to 7 p.m.; one dispatcher was used during non-peak periods. This allowed for $52,000 in annual cost savings for the company while enhancing customer service and response time (the previous policy used two dispatchers at all times). The final improvement involved hiring one additional P2 plumber as the P2 plumber category was overutilized (as demonstrated previously in Figure 19.3).

19.2.5 Control

To monitor and maintain the improvements made, the Lean Six Sigma team implemented several tracking tools that are run on a smartphone platform. Plumbers, dispatchers, and managers were all given smartphones to track jobs and monitor effort and travel required on a per-job basis. Whenever the plumber's utilization dropped below 80% (out of control situation), an instant message was automatically sent to the dispatcher and manager. In addition the dispatch and management team monitored the daily average inflow volume of service requests. The ticketing system data were leveraged in a macro-enabled spreadsheet for control chart analysis on a nightly basis for each customer process or combination of customer processes. These control charts and dashboards were integrated with daily management meetings to discuss challenges, provide required help, and plan future actions.

20

Case Study 9: Service—15-Minute Oil Change

20.1 Introduction

The company studied was a 15-minute automobile oil change provider operating in Toledo, Ohio. The company employed five service technicians and one supervisor and operated seven days per week from 7 a.m. to 7 p.m. Recently, a new quick oil change business opened in the area and the company was beginning to drop sales, suffer lower customer satisfaction, higher oil change times, and high employee turnover. The purposes of the study were to understand the cause of the issues, quantify the issues, identify the cause, and improve productivity and customer service.

20.2 Application of Lean DMAIC Process

20.2.1 Define

For the first phase of the project, the Lean Six Sigma team, consisting of the manager, the lead technician, and an outside consultant reviewed the current operations and financial records. The team found that, based on online surveys, customer satisfaction was dropping, revenue was down 15% from the previous year, and the time to complete oil changes was displaying higher completion times and variances. The goals of the Lean Six Sigma project were to increase productivity by reducing and stabilizing oil change times to 15 minutes or less, increasing customer satisfaction by 20%, and reducing employee turnover by 20%.

20.2.2 Measure

The goals of the measurement phase were to quantify and baseline the current performance of the quick oil change operations. Specifically, the Lean Six Sigma team was interested in quantifying:

- The number of oil changes per day
- Peak demand periods versus staffing levels
- Value-added versus non-value-added activities
- Process times, variations, delays, and waste in the process
- Customer satisfaction
- Employee attitudes and satisfaction

The team reviewed computer and paper records to determine that the facility completed 22 oil changes per day on average with a standard deviation of 1.8. Figure 20.1 displays a graph that compares customer arrivals to staffing levels.

As displayed in Figure 20.1, the staffing levels were consistent whereas customer demand varied significantly. Next, the team conducted a time study on a typical oil change process and a delay study for the period of one shift. The results are displayed in Tables 20.1 and 20.2, respectively.

As shown in Table 20.1, the oil change process required 19.1 minutes, 4.1 minutes over the "15-minute oil change" promise by the company. Additionally, as displayed in Table 20.2, the process experienced approximately 11.6% in avoidable delays (over 83 minutes on a 12-hour work shift). Finally, the Lean Six Sigma consultant measured customer and employee satisfaction using a short five-question survey. The customer survey measured satisfaction on a scale of 1 to 5 where 1 = very dissatisfied, 2 = dissatisfied, 3 = neutral, 4 = satisfied, 5 = very satisfied. The results of the survey indicated that the customers were dissatisfied with the time required to complete the oil changes, wait times during peak demand periods, and what appeared to be a lack of urgency by technicians to complete oil changes in a timely fashion. Otherwise, the customers were satisfied with the process. The employee survey indicated that the employees felt overworked during peak demand

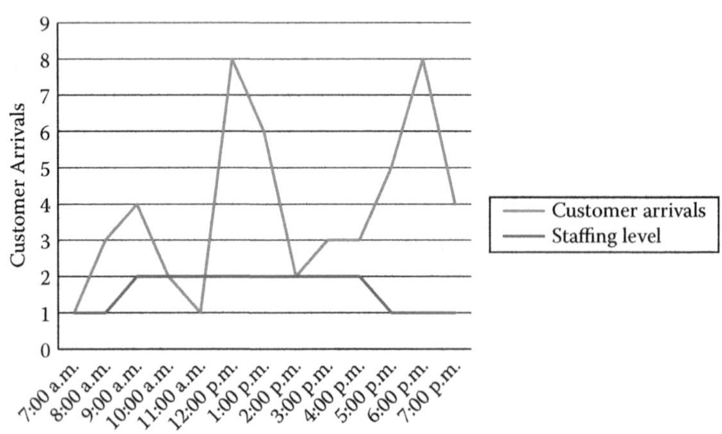

FIGURE 20.1
Customer demand versus staffing levels.

TABLE 20.1

Time Study Results

Activity	Time Required (min.)
Greet customer	1.2
Log customer information in computer	0.8
Gather parts and tools	1.8
Drain oil	2.9
Replace oil	3.1
Replenish fluids	2.8
Wash car windows	1.0
Discuss results with customer	1.8
Take payment	2.0
Log customer information in computer	0.9
Clean work area for next customer	0.8
TOTAL	**19.1**

TABLE 20.2

Delay Study Results

Cause of Delay	Time Lost (min.)
Unable to find filters/parts	22.3
Computer error/logging delays	8.1
Unauthorized employee break	9.5
Customer conversation	24.3
New customer data entry	19.6
Total delay	**83.8**
Time observed	**720**
% Delay	**11.6%**

periods and felt additional help were needed. Two of the technicians indicated they were not properly trained to use the computer and lacked skills to enter new customer information efficiently.

20.2.3 Analyze

For the analysis stage, the Lean Six Sigma team reviewed the data to identify trends and areas to target for improvement. From an initial review of the data, the team identified:

- Staffing levels were not matched to customer arrival demand (both over- and underutilization at various times).
- Significant delays were present and the process was not capable of achieving the 15-minute oil change promise as advertised.
- Employee morale was low.

As previously displayed in Figure 20.1, staffing levels were too high (underutilization) between 7–11 and 2–5 and too low (overutilization) between 11–2 and 5–7. Figure 20.2 displays an average customer wait-time graph during hourly time intervals. The wait times were recorded over a two-week period. Figure 20.3 displays the average customer process time (wait time and service time) by hourly time intervals.

As displayed in Figure 20.3, during peak times, the process is not able to meet the 15-minute oil change time. This was also affecting employee

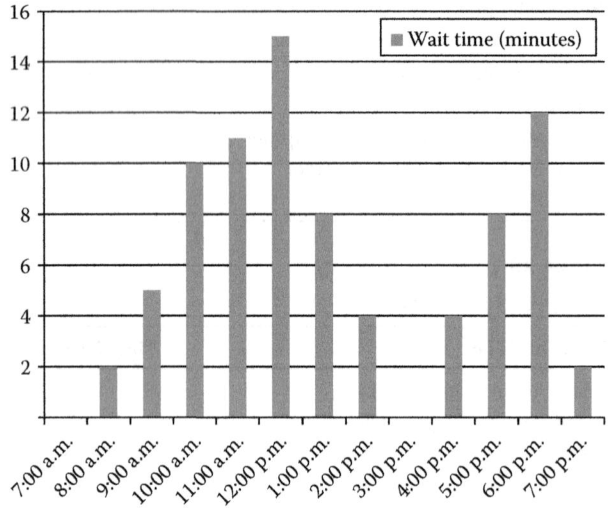

FIGURE 20.2
Average customer wait times.

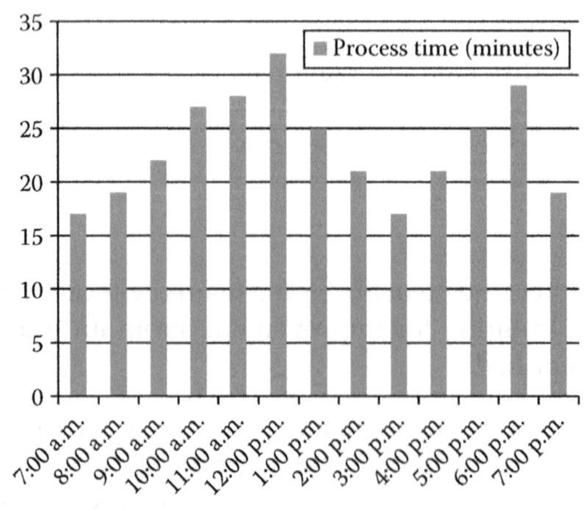

FIGURE 20.3
Average customer process times.

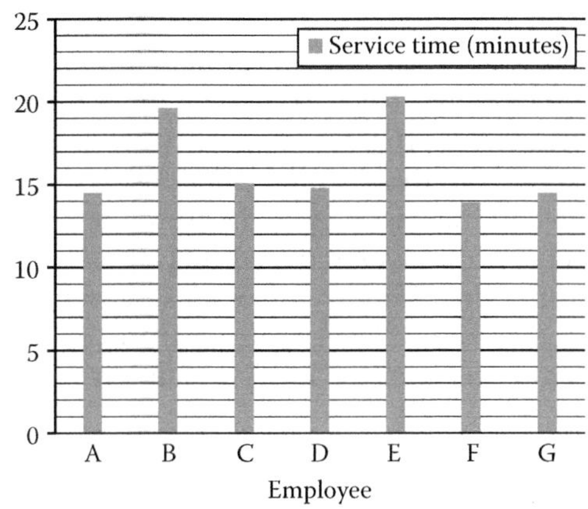

FIGURE 20.4
Average oil change time per employee.

morale, as employees would feel overworked or underworked. Also, two "problem" employees were identified who were requiring significantly longer than other employees to complete oil changes. Figure 20.4 displays the average oil change time by employee.

As shown in Figure 20.4, the two problem employees (Employees B and E) were requiring over 30% longer (4 to 5 minutes longer) to complete oil changes versus the other employees who were meeting the 15-minute oil change guarantee. Digging deeper into the employee morale issue, the Lean Six Sigma consultant identified that the two problem employees were dragging down morale for the entire work team. The challenges were to increase/reduce staffing levels to match customer demand, reduce process delays, and increase employee morale.

20.2.4 Improve

To capitalize on the findings from the analysis and measurement phases, the Lean Six Sigma team implemented several improvement initiatives. First, the two problem employees were met with separately to discuss performance expectations, their current performance, and recommendations for changes. The management discovered that both employees were not computer literate and the data entry process was the cause of major delay for both. To improve and reduce the data entry times, both employees were given additional computer training and a driver's license scanner was added to quickly scan/upload driver information and addresses into the database. The company also created a more fun atmosphere for employees and customers by creating a NASCAR "pits" theme to help meet the 15-minute oil change guarantee. A large timer was added to display the start and end times of the oil change for customers and employees and if an oil change took longer than 15 minutes, the customer was given a 20% off the next oil change coupon. Finally, the company created a smartphone app to allow customers to schedule oil changes and view the current wait times for an oil change. From a staffing perspective two to three employees were scheduled during peak demand times to keep wait times down. As a result of the changes, within six months, revenue had increased by 18.5% from the previous year, and the time to complete oil changes was reduced to 14.3 minutes with a standard deviation of 1.1 minutes. In addition, customer satisfaction as determined by customer survey increased to 4.2 (very satisfied) and employee turnover was zero during the six months after the implementation of improvements.

20.2.5 Control

To monitor and maintain the improvements made, the Lean Six Sigma team implemented several tracking tools that are run on a smartphone platform. Managers were all given smartphones to track jobs and monitor oil change times. Daily charts were posted in employee break areas to display average oil change times for the group and the employee with the lowest time was given an award and gift card.

21

Case Study 10: Education

21.1 Introduction

This case study took place at a community college in the midwestern United States that was experiencing a high level of student complaints related to student services and advising. These issues were beginning to lead to low student satisfaction and decreasing enrollment. The college studied was a state-funded institution that enrolled 12,000 students and operated on the semester system. At the request of the provost of the college, the Student Services Department was requested to embark on a rapid analysis and improvement process. The faculty in the Business Department was contacted and the team determined that Lean Six Sigma would be a good fit for the college.

21.2 Application of Lean DMAIC Process

21.2.1 Define

In defining the problem, the Lean Six Sigma team first recognized that the student was both a customer and product of the institution. As a result, the team determined that most of the Student Services Department's activities related to the students as customers of the college, whereas the educational activities related more to the students as products. For the purpose of improving the Student Service Department, the students were identified as the primary customers. After several meetings with student focus groups and college administration, the Lean Six Sigma team identified low student satisfaction, long service wait times, sometimes unprofessional/rude student service, and decreasing enrollment as the primary issues to address. The goals of the team were to baseline student satisfaction levels and improve the overall rating by 15% within one year, decrease student advising wait times by 25%, and increase enrollment by 5% within one year.

21.2.2 Measure

The goals of the measurement phase were to baseline the current performance of the system and track student satisfaction. First a student support services satisfaction survey was given to students to complete. The survey was announced via college e-mail and completed electronically. To increase participation, three computer tablets were raffled in a drawing for those who completed the survey. The results of the survey are presented in Table 21.1.

Scale:

- 1—Very dissatisfied
- 2—Somewhat dissatisfied
- 3—Neutral
- 4—Somewhat satisfied
- 5—Very satisfied

In addition to the survey, time studies were conducted on student wait times for walk-in advising, student advising service times, lead time to schedule an appointment, and response times to e-mails and phone messages. The results of the time studies are presented in Table 21.2.

TABLE 21.1

Student Satisfaction Survey Results

Survey Question	Average
How satisfied have you been at the campus	4.2
Organized out of class experiences	5.1
Informal out of class experiences	4.9
Sense of belonging at the campus	4.8
Safety and security at the campus	4.7
Orientation program	2.8
Quality of major courses on campus	3.9
Quality of general courses on campus	3.1
Availability of courses to make progress toward your degree	4.5
Registration process	2.4
Advising resources available online	1.6
Accessibility of your academic adviser	1.4
Information and referrals provided by your academic adviser	1.7
Overall quality of academic advising	2.0
If you were starting college again, would you choose this college?	58% Yes

TABLE 21.2
Time Study Results

Activity	Time Required (min.)
Arrive in Student Service Office	0.5
Wait to meet adviser	6.9
Greeting with adviser	0.7
Adviser gets student folder from file area	0.8
Discussion with student	3.4
Problem or issue resolution	5.1
Exit from adviser	0.8
Adviser documents meetings in student folder	3.5
Adviser files student folder	0.8
Total	**22.5**

21.2.3 Analyze

Upon analyzing the data collected during the measurement phase, the Lean Six Sigma team identified several critical to quality issues associated with the Student Services Department:

- Students were not satisfied with the availability of resources.
- Students were not satisfied with the availability of advisers.
- Students were not satisfied with information/referrals from advisers.

In combination with the time study data, the average wait time to see an adviser was approximately 7 minutes with an overall wait/service time of 22.5 minutes. Based on these findings the Lean Six Sigma team formed a focus group of students to explore more deeply the low satisfaction ratings. The findings from the focus group indicated:

- Students were not satisfied with long wait times.
- Students would prefer advisers with experience in their respective majors as opposed to the general non-major–specific advisers currently employed.
- The time for advisers to return phone calls or respond to e-mails seemed exceptionally high and some students reported never receiving responses.
- Students would prefer early evening hours for advising; currently advisers are available from 9 a.m. to 3 p.m. Monday through Friday. Students would like hours after 5 p.m. to better accommodate school and work schedules.

21.2.4 Improve

To improve the student service processes the Lean Six Sigma team implemented several improvements based on the data and focus group results:

- A faculty adviser was added as another resource for students for each major to provide perspective on technical aspects and job prospects.
- Text and online chat options were added for advising.
- An adviser was removed from meeting with students and dedicated to answering e-mails, phones, texts, and scheduling appointments and online chats.
- Advising hours were extended into the evening until 7:30 p.m. on Monday through Thursday.
- Advisers were provided with "customer service" training to improve student interactions and friendliness.
- An online database system was created to store student records electronically. This allowed all advisers to have immediate access to students' real-time files from any computer. It also reduced filing time and notetaking to improve efficiency.

21.2.5 Control

To monitor and maintain the improvements made, the Lean Six Sigma team implemented several tracking tools and surveys to measure student satisfaction and service times. The surveys were administered via e-mail twice per semester and included a raffle for a tablet. A "secret shopper" program was also implemented to measure adviser service levels and friendliness. The results of the program were shared with advisers to help improve performance.

22

Case Study 11: Nonprofit and Government

22.1 Introduction

This case study analyzed a Job and Family Services office in the midwestern United States; it is a nonprofit and government-supported group that employs 18 full-time employees and assists over 1,250 families per year. Typically, Lean Six Sigma is applied to reduce costs and boost profits for for-profit groups, but it may also be applied to nonprofits to improve service levels to its constituents and the taxpayers. The Job and Family Services office studied provided services such as temporary cash assistance, food stamps, Medicaid, subsidized child care, adult protection, disability assistance, and prevention, retention, and contingency services. The Division of Child Support establishes parentage, enforces support orders, and collects and disburses child support in order to ensure that families receive the financial support to which they are entitled. The organization began to experience long wait lists and they were overwhelmed and busy. The management team felt that hiring a new staff member would have taken at least six months of training, so improving the system and process to reduce the wait time for clients was the most viable option.

22.2 Application of Lean DMAIC Process

22.2.1 Define

To assist with the Lean Six Sigma process, the organization hired a Lean Six Sigma consultant using funds from a state grant to assist in improving processes. The primary issue that the organization was experiencing was related to long wait lists to schedule appointments. To compound the issue the organization had a large percentage of no-shows for scheduled appointments. The goal of the study was to reduce the wait-list time by 25% and decrease the no-show rate by 50%.

22.2.2 Measure

The Lean Six Sigma team had a general understanding of wait times and service issues, but lacked hard data. The first step was to collect the required hard data and to determine baseline measurements. As mentioned, client no-show rates were relatively high, so the team would overbook appointments to account for these no-shows; this appeared to be leading to an unstable and unpredictable system with long delays and lower customer satisfaction. The Lean Six Sigma team conducted six simultaneous studies over a four-week period:

- Time studies of the process to serve a client (Table 22.1)
- A delay study to determine non-value-added activities from the customers' perspective (Table 22.2)
- A study of client arrival, service, and no-show rates, including the number of customers served per day by each employee (Figure 22.1)
- Lead times to schedule an appointment (Table 22.3)
- Client satisfaction survey (Table 22.4)
- Fishbone analysis for long lead times and no-shows (Figure 22.2)

TABLE 22.1

Process Time Study Data

Activity	Time Required (min.)
Greet client	1.8
Gather customer information/complete forms	3.5
Discuss client issues/concerns	6.8
Determine resolutions and available support	5.2
Close with client	1.4
Record meetings notes in client file	1.6
Total	**20.3**

TABLE 22.2

Delay Study Data

Cause of Delay	Time Lost (min.)
Client late arrival/no-show	41.3
Searching for client file	32.7
Employee breaks	12.1
Conversations with other employees	10.8
Total Delay	**96.9**
Time Observed	**540**
% Delay	**17.9%**

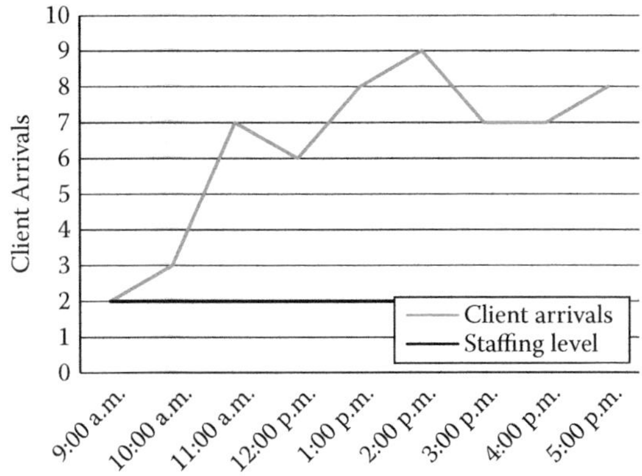

FIGURE 22.1
Service and arrival rate analysis.

22.2.3 Analyze

Based on the baseline data collected during the measurement phase, the Lean Six Sigma team identified several issues related to wait times, lead times, and client satisfaction:

- Overall, clients were satisfied with advising and contact with advisers based on survey results.
- Overall, clients were dissatisfied with wait times and lead times to schedule appointments based on survey results and time study data.

In combination with the time study data and lead time data, the average lead time to see an adviser was 8.3 days with a standard deviation of 2.22 days. The average client meeting was 20.3 minutes based on time study data; of that time 17.9% was non-value-added. In addition, the staffing versus client arrival rate indicated higher client arrival rates later in the day and a higher disconnect between the staffing level and number of clients needing to be served. Finally, the no-show rates for clients during the four-week study period was 34.6% for appointments and a late-arrival rate of 25.4% for the clients who appeared for scheduled meetings.

22.2.4 Improve

To improve the Job and Family Service processes, the Lean Six Sigma team implemented several improvements based on the data and analysis phases:

- A precheck in text, e-mail, or phone call system was added for clients the morning of appointments to minimize no-shows and late arrivals. Clients were required to check in at least 30 minutes prior

TABLE 22.3

Lead Time to Schedule an Appointment

Client	Lead Time (Days)
1	4
2	5
3	8
4	9
5	8
6	7
7	9
8	10
9	7
10	12
11	6
12	9
13	8
14	9
15	7
16	6
17	10
18	14
19	6
20	7
21	8
22	10
23	12
24	8
25	9
Average	**8.32**
Standard Deviation	**2.22**

to their appointment or they would lose it. If a client missed two or more scheduled appointments, he or she would be required to attend a 30-minute time management seminar before another appointment could be scheduled.

- An online database system was created to store client records electronically. This allowed for all advisers to have immediate access to clients' real-time files from any computer. It also reduced filing time and notetaking to improve efficiency.
- One more adviser was added after 12 p.m. to better match client demand.

As a result of the changes, the no-show rate dropped from 34.6% to 12.8%. A second project focused on reducing the waiting list, with wait times from

TABLE 22.4

Client Satisfaction Survey

Survey Question	Average
How satisfied have you been with your experience?	3.1
How satisfied have you been with wait times?	1.6
How satisfied have you been with lead time to schedule an appointment?	1.9
Quality of services provided	4.6
Accessibility of your adviser	1.8
Information and referrals provided by your adviser	4.3
Overall quality of advising	4.8

Scale:
1—Very dissatisfied
2—Somewhat dissatisfied
3—Neutral
4—Somewhat satisfied
5—Very satisfied

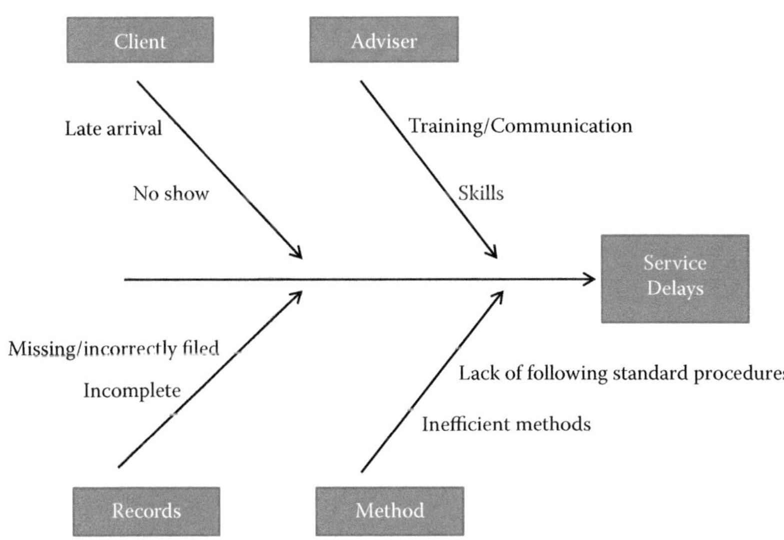

FIGURE 22.2
Fishbone analysis.

counseling applications dropping from 8.3 days to 1.5 days. The organization went from 300 children on its waiting list to less than 50, while reducing the total cycle time from 65 days down to about 25 days from the time of the first call until the first scheduled advising session. The changes also resulted in increased referrals from other care providers, who knew their clients would now receive speedier service.

22.2.5 Control

To maintain results, no-show, late-arrival, and service time reports were automatically generated once per week. The administrative and advising team would review the results and determine any needed changes. Clients were also sent text or e-mail updates regarding scheduled appointments and next steps. A client postsurvey indicated a satisfaction increase for wait times from 1.6 to 4.1.

Index

Note: Page numbers ending in "f" refer to figures. Page numbers ending in "t" refer to tables.

M